EMERGING TRENDS OF RESEARCH IN CHEMICAL SCIENCES

Qualitative and Quantitative Studies and Applications

EMERGING TRENDS OF RESEARCH IN CHEMICAL SCIENCES

Qualitative and Quantitative Studies and Applications

Edited by
Tanmoy Chakraborty, PhD
Shalini Chaudhary, PhD

First edition published 2022

Apple Academic Press Inc.
1265 Goldenrod Circle, NE,
Palm Bay, FL 32905 USA

4164 Lakeshore Road, Burlington,
ON, L7L 1A4 Canada

CRC Press
6000 Broken Sound Parkway NW,
Suite 300, Boca Raton, FL 33487-2742 USA

2 Park Square, Milton Park,
Abingdon, Oxon, OX14 4RN UK

© 2022 Apple Academic Press, Inc.

Apple Academic Press exclusively co-publishes with CRC Press, an imprint of Taylor & Francis Group, LLC

Library and Archives Canada Cataloguing in Publication

Title: Emerging trends of research in chemical sciences : qualitative and quantitative studies and applications / edited by Tanmoy Chakraborty, PhD, Shalini Chaudhary, PhD.

Names: Chakraborty, Tanmoy, editor. | Chaudhary, Shalini, editor.

Description: First edition. | Includes bibliographical references and index.

Identifiers: Canadiana (print) 20210122455 | Canadiana (ebook) 20210122471 | ISBN 9781771889735 (hardcover) | ISBN 9781774638286 (softcover) | ISBN 9781003129929 (ebook)

Subjects: LCSH: Chemistry—Research.

Classification: LCC QD40 .E44 2021 | DDC 540.72—dc23

Library of Congress Cataloging-in-Publication Data

Names: Chakraborty, Tanmoy, editor. | Chaudhary, Shalini, editor.

Title: Emerging trends of research in chemical sciences : qualitative and quantitative studies and applications / edited by Tanmoy Chakraborty, PhDShalini Chaudhary, PhD.

Description: First edition. | Palm Bay, FL : Apple Academic Press, 2021. | Includes bibliographical references and index. | Summary: "Emerging Trends of Research in Chemical Sciences: Qualitative and Quantitative Studies and Applications is compilation of the research being done by scientists from various disciplines of chemistry at various universities across the globe. This new volume provides a wealth of practical experience and research on new methodologies and important applications in chemical science. It also includes presentations on small-scale new drug design related projects that have potential applications in several disciplines of chemistry and in drug development. In this book, contributions range from new methods to novel applications of existing methods to enhance understanding of the material and/or structural behavior of new and advanced systems. Topics cover computational methods in chemical sciences and electrochemical investigations, studies of some of physico-chemical properties of several important novel macrocyclic ligands, the use of lanthanide-ions doped nanomaterials, quantitative estimation of heavy metals, sustainable, efficient and green promoter for the synthesis of some heterocyclic compounds, and much more"-- Provided by publisher.

Identifiers: LCCN 2021003361 (print) | LCCN 2021003362 (ebook) | ISBN 9781771889735 (hardcover) | ISBN 9781774638286 (paperback) | ISBN 9781003129929 (ebook)

Subjects: LCSH: Chemistry--Research. | Chemistry--Methodology.

Classification: LCC QD40 .E37 2021 (print) | LCC QD40 (ebook) | DDC 540.72--dc23

LC record available at https://lccn.loc.gov/2021003361

LC ebook record available at https://lccn.loc.gov/2021003362

ISBN: 978-1-77188-973-5 (hbk)
ISBN: 978-1-00312-992-9 (ebk)
ISBN: 978-1-77463-828-6 (pbk)

About the Editors

Tanmoy Chakraborty, PhD

Associate Professor, Department of Chemistry and Biochemistry, School of Basic Sciences and Research, Sharda University, Knowledge Park III, Greater Noida 201310, Uttar Pradesh, India.

Tanmoy Chakraborty, PhD, is now working as Associate Professor, Department of Chemistry & Biochemistry, School of Basic Sciences & Research, Sharda University, India. He has been working in the challenging field of computational and theoretical chemistry for the last 15 years. He has completed his PhD from the University of Kalyani, West Bengal, India, in the field of application of QSAR/QSPR methodology in the bioactive molecules. He has published more than 50 research papers in peer-reviewed international journals with high impact factors. Dr. Chakraborty served as an international editorial board member of the *International Journal of Chemoinformatics and Chemical Engineering*, IGI Global, USA. He has edited / written several books / book chapters with renowned international publishing house. He has received around INR 45 Lakh research grant from Department of Science and Technology - Science and Engineering Research Board, Government of India. Dr Chakraborty is reviewer of high repute journals of ACS, Springer, Elsevier, Wiley. Currently he is performing the role of Lead Guest Editor of the special issue (dedicated to celebrate 80th Birthday of Prof (Dr) Ramon Carbo Dorca) of the Journal 'Theoretical Chemistry Accounts.' Dr. Tanmoy Chakraborty is the recipient of prestigious Paromeswar Mallik Smawarak Padak, from Hooghly Mohsin College, Chinsurah (University of Burdwan), in 2002.

Shalini Chaudhary, PhD

*Assistant Professor and Head of the Department of Chemistry,
Alankar P. G. Girls College, Jaipur 302012, Rajasthan, India.*

Shalini Chaudhary, PhD, one of the editors of this book is currently Assistant Professor, Head of the Department of Chemistry at Alankar P.G. Girls College, Jaipur, India, since 15 years. She has completed her PhD in the field of computational and theoretical chemistry from Manipal University, Jaipur. She is enthusiastically involved in the field of computational and physical chemistry. She has published several papers and book chapters in a number of renowned journals and books. She has also attended various national and international conferences.

Contents

Contributors

Sevim Akyuz
Physics Department, Science and Letters Faculty, Istanbul Kultur University, Atakoy Campus, Bakirkoy, Istanbul 34156, Turkey

Sunita Bhagat
Department of Chemistry, ARSD College, University of Delhi, Dhaula Kuan, New Delhi 110021, India

Sangeeta Bhargava
Department of Chemistry, Centre of Advanced Studies, University of Rajasthan, Jaipur 302004, India

Sefa Celik
Physics Department, Science Faculty, Istanbul University, Vezneciler, Istanbul 34134, Turkey

Anita Choudhary
Department of Chemistry, Centre of Advanced Studies, University of Rajasthan, Jaipur 302004, India

Nighat Fahmi
Department of Chemistry, University of Rajasthan, Jaipur 302004, India

Nidhi Gaur
Department of Chemistry, University of Rajasthan, Jaipur, Rajasthan, India

Neelima Gupta
Centre for Advanced Study, Department of Chemistry, University of Rajasthan, Jaipur, India

Krishna Kumar Jhankal
Electrochemical Research Laboratory, Department of Chemistry, University of Rajasthan, Jaipur 302004, Rajasthan, India

Varsha Jakhar
Electrochemical Research Laboratory, Department of Chemistry, University of Rajasthan, Jaipur 302004, Rajasthan, India

Romila Karnawat
Govt. SCRS College, Sawaimadhopur, Rajasthan, India

Gunjan Kashyap
Department of Chemistry, ARSD College, University of Delhi, Dhaula Kuan, New Delhi 110021, India

Serda Kecel-Gunduz
Physics Department, Science Faculty, Istanbul University, Vezneciler, Istanbul 34134, Turkey

Amit Kumar
Department of Chemistry, ARSD College, University of Delhi, Dhaula Kuan, New Delhi 110021, India

Ginni Kumawat
Forensic Science Laboratory, Jaipur, Rajasthan, India

Sandeep Nigam
Chemistry Division, Bhabha Atomic Research Centre Mumbai, Mumbai 400085, India

Aysen E. Ozel
Physics Department, Science Faculty, Istanbul University, Vezneciler, Istanbul 34134, Turkey

Prateek Pandya
Amity Institute of Forensic Sciences, Amity University, Noida, India

Chandresh Kumar Rastogi
Chemistry Division, Bhabha Atomic Research Centre Mumbai, Mumbai 400085, India

Deepti Rathore
Department of Chemistry, Centre of Advanced Studies, University of Rajasthan, Jaipur 302004, India

D. K. Sharma
Electrochemical Research Laboratory, Department of Chemistry, University of Rajasthan, Jaipur 302004, Rajasthan, India

I. K. Sharma
Department of Chemistry, University of Rajasthan, Jaipur, Rajasthan, India

Naveen Sharma
Department of Chemistry, University of Rajasthan, Jaipur, 302004, India.

Nutan Sharma
Department of Chemistry, ARSD College, University of Delhi, Dhaula Kuan, New Delhi 110021, India

Pankaj Sharma
Department of Chemistry, ARSD College, University of Delhi, Dhaula Kuan, New Delhi 110021, India

R. V. Singh
Department of Chemistry, University of Rajasthan, Jaipur 302004, India

Kshipra Soni
Department of Chemistry, University of Rajasthan, Jaipur 302004, India

V. Sudarsan
Chemistry Division, Bhabha Atomic Research Centre Mumbai, Mumbai 400085, India

Monika Upadhyay
Department of Chemistry, University of Rajasthan, Jaipur 302004, India

P. S. Verma
Department of Chemistry, University of Rajasthan, Jaipur, Rajasthan, India

Abbreviations

$A21$	asymmetry parameter
AAS	atomic absorption spectroscopy
ATSDR	Agency for Toxic Substances and Disease Registry
BR	Britton–Robinson
CN	coordination number
CTAB	cetyltrimethyl ammonium bromide
CV	cyclic voltammetry
EDA	ethylenediamine
EDTA	ethylenediaminetetraacetic acid
EPA	Environmental Protection Agency
GCE	glassy carbon electrode
HFIP	hexafluoroisopropanol
HRTEM	high-resolution transmission electron micrographs
LbL	layer-by-layer
LEDs	light-emitting diodes
LOD	limit of detection
LOQ	limit of quantization
LMCT	ligand to metal charge transfer
Ln	lanthanides
Ln-NPs	lanthanide-doped nanoparticles
MC	Monte Carlo
MD	molecular dynamic
MDR	multidrug resistance
NFPY	N-fluoropyridinium tetrafluoroborate
NFSI	N-fluorobenzenesulfonimide
NPs	nanoparticles
OA	oleic acid
OAM	oleylamine
ODE	1-Octadecene
PDP	plasma display panel
PES	potential energy surface
PET	positron emission tomography
PPL	polypropylene

PTFE	polytetrafluoroethylene
PVP	polyvinylpyrrolidone
QDs	quantum dots
QY	quantum yield
SPSS	Statistical Package for the Social Sciences
SW–ASV	square wave anodic stripping voltammetry
TLC	thin-layer chromatography
TM	transition metal
TMS	tetramethylsilane
TOPO	tri-*n*-octylphosphine oxide
UCNPs	up-converting nanoparticles
VUV	vacuum ultraviolet

Preface

This book, *Emerging Trends of Research in Chemical Sciences: Qualitative and Quantitative Studies and Applications,* has been successfully edited by Dr. Tanmoy Chakarborty and Dr. Shalini Chaudhary, who are budding scientists and experts in the field of computational and theoretical chemistry.

The idea for this book initially stemmed from the desire to provide an opportunity for researchers from various disciplines of chemistry to present their work on a single platform. It is a compilation of the research being done by scientists at various universities across the globe.

In this new book, we aim to cover the important applications of computational chemistry in diverse fields. It targets a wide audience ranging from the fields of heterocyclic chemistry, green chemistry, computational chemistry, quantum chemistry to medicinal chemistry.

This new volume provides a wealth of practical experience and research on new methodologies and important applications in chemical science. It also includes small-scale new drug design-related projects that have potential applications in several disciplines of chemistry and drug industry.

In this book, contributions range from new methods to novel applications of existing methods to gain understanding of the material and/or structural behavior of new and advanced systems.

The fast growing field of nanotechnology uses a variety of modern methods that give scientists the ability to design multifunctional materials and provide improved products. Understanding the world of nanotechnology is critically important because technology based on different research promises to be hugely important economically.

Topics cover computational methods in chemical sciences and electrochemical investigations, studies of some of physicochemical properties of several important novel macrocyclic ligands, the use of lanthanide-ions doped nanomaterials, quantitative estimation of heavy metals, and sustainable, efficient, and green promoters for the synthesis of some heterocyclic compounds and much more.

The significant feature of this addition is that the chapters have been included from the various fields of the research of chemical sciences. We hold this golden opportunity to write, this preface of the present book and it is to state that the seed of this book was sown by Professor Tanmay Chakraborty. The present fruit is the sprout of their books and their other celebrated books in the series "Innovation in Computational Chemistry."

We are deeply indebted to Apple Academic Press who gave us an opportunity to prepare this book in chemistry as needed by the changed scenario of the research.

The present book is meant for young researchers at universities; it covers the various field of research of chemical sciences and innovation recently. We intensely believe that science is not a plethora of dry and abstract facts on the platform of the laboratory; these are the laws of nature to be really understood and applied for the benefit of mankind.

CHAPTER 1

Electrochemical Investigations of Antibiotic Drug Linezolid in Pharmaceuticals and Spiked Human Urine by Stripping Voltammetric Techniques

KRISHNA KUMAR JHANKAL, VARSHA JAKHAR, and D. K. SHARMA*

Electrochemical Research Laboratory, Department of Chemistry, University of Rajasthan, Jaipur 302004, Rajasthan, India

Corresponding author. E-mail: sharmadkuor@gmail.com

ABSTRACT

In present study, a sensitive and fast voltammetric method was developed for the trace analysis of antibacterial drug linezolid in pharmaceuticals and in human plasma at glassy carbon electrode (GCE). The oxidation of linezolid gave reversible peak in Britton–Robinson buffer at GCE in the potential range of 1.0–1.1 V versus Ag/AgCl/KCl reference electrode. Since the process was diffusion controlled, a linear response was obtained between 1.0×10^{-7} and 5.0×10^{-5} M in aqueous media. The proposed method was applied for the determination of linezolid in human urine without any time-consuming extraction, separation, and adsorption steps.

1.1 INTRODUCTION

Linezolid, [N-{[(5S)-3-[3-fluoro-4-(morpholin-4-yl)phenyl]-2-oxo-1,3-oxa-zolidin-5-yl]methyl}acetamide] (Fig. 1.1) is a synthetic antibiotic that is used for the treatment of serious infections caused by Gram-positive bacteria that are resistant to several other antibiotics. A member of the oxazolidinone class of drugs, linezolid is active against most Gram-positive bacteria that cause several types of diseases that include streptococci, vancomycin-resistant enterococci, and methicillin-resistant *Staphylococcus aureus*.[1,2]

Linezolid mainly indicates infections of the skin and soft tissues and pneumonia, although its off-label use for a variety of other infections is becoming popular day by day. As a protein synthesis inhibitor, it inhibits bacteria to grow by disrupting their production of proteins that is, it is a bacteriostatic agent, not bactericidal.[3–5]

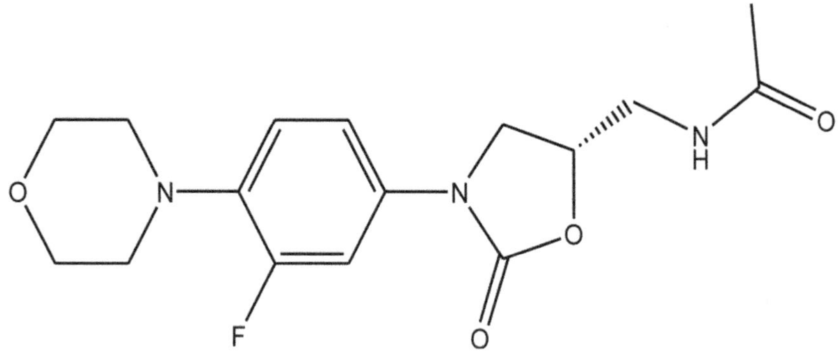

FIGURE 1.1   Chemical structure of linezolid.

Among many bacteriostatic agents, the mechanistic action of linezolid appears to be unique as it specifically blocks the initiation of protein production. In both the popular press and the scientific literature, linezolid has been called a "reserve antibiotic" that should be used sparingly so that it will remain effective as a drug of last resort against potentially intractable infections,[6] and linezolid appears to be as safe and effective for use in children and newborns as it is in adults. Linezolid is official in Indian Pharmacopoeia.[7]

The analytical methods in previous literature have reported the estimation of linezolid in pharmaceutical dosage forms: spectrophotometry,[8] ultraviolet (UV) spectroscopy,[9] liquid chromatography–UV spectroscopy,[10] high performance liquid chromatography (HPLC)–UV spectroscopy,[11] HPLC,[12] RP-HPLC,[13] and HPTLC[14] methods.

Electrochemical methods have proved to be sensitive and reliable for the determination of numerous electroactive drug components.[15–25] Under some circumstances, electrochemical methods can offer optimal solutions for drug analysis. The electroanalytical techniques proved to be useful both for the analysis of pharmaceutical dosage forms and determination of drugs in biological fluids.[26–33]

In the present study, electrochemical behavior of linezolid at glassy carbon electrode (GCE) was investigated using cyclic voltammetry (CV), differential pulse and square wave anodic stripping voltammetry (DP–ASV and SW–ASV).

1.2 EXPERIMENTAL PART

1.2.1 REAGENTS AND MATERIALS

Linezolid was purchased under the trade name Lizoforce® from Mankind Pharma Ltd. India and was used without further purification. A stock standard solution of bulk linezolid (1×10^{-3} M) was prepared in methanol–water (70:30, %v/v) and stored at 4°C until assay. The content of the flask was sonicated for about 10 min. The working solutions (1×10^{-6} to 1×10^{-4} M) were prepared daily by appropriate dilution of the standard solution of bulk linezolid just before use. A series of Britton–Robinson (BR) buffer of pH values 3–11 was prepared and used as a supporting electrolyte.

1.2.2 APPARATUS

Model 1230A (SR 400) electrochemical analyzer (CHI Instrument, United States) was employed for electrochemical techniques, with a totally auto-mated attached to a computer with proper CHI 100W version 2.3 software for total control of the experiments and data acquisition and treatment. A three-electrode cell system was used with activated GCE as working electrode, Ag/AgCl (3 M KCl) as the reference electrode, and a platinum wire as the auxiliary electrode. A digital pH meter (CHINO-DB-1011) was used for measuring the pH values of the investigated solutions.

1.2.3 PRETREATMENT OF THE GLASSY CARBON ELECTRODE

The GCE was polished with 0.5 µm alumina powder on a polishing cloth prior to each electrochemical measurement.

Then it was thoroughly rinsed with methanol and double distilled water and gently dried with a tissue paper.

1.2.4 GENERAL ANALYTICAL PROCEDURE

10 mL of the total solution containing BR buffer and the appropriate concentration of the linezolid were transferred into the electrochemical cell, through which a pure deoxygenated nitrogen stream was passed for 10 min to remove the oxygen gas before measurements. Electrochemical pretreatment was always performed in the same solution, in which the measurement was subsequently carried out. The accumulation of linezolid at the working electrode was carried out for a selected time, while the solution was stirred at 1000 rpm. The stirring was then stopped, and after a rest period voltammetric processes were initiated in the anodic direction over the range of 0.4–1.2 V versus Ag/AgCl/KCl reference electrode at room temperature.

1.2.5 ANALYSIS OF SPIKED HUMAN URINE SAMPLES

Drug-free human urine, obtained from healthy volunteers, was centrifuged (4000 rpm) for 30 min at room temperature, and separated samples were stored frozen until assay. An aliquot of urine sample (1.0 mL each) was fortified with various concentrations of linezolid (1×10^{-7} to 1×10^{-5} M) in centrifuge tubes then each was mixed with a 1.0 mL volume acetonitrile to denature and precipitate proteins. After vortexing for 30 s, the mixture was then centrifuged for 10 min at 4000 rpm in order to eliminate protein residues. Appropriate volumes of this sample were transferred into the voltammetric cell and diluted up to the volume with BR buffer and subsequently analyzed according to the recommended in the general analytical procedure.

1.3 RESULTS AND DISCUSSION

The electrochemical behavior of linezolid was studied by CV. DP–ASV and SW–ASV technique were developed for the determination of linezolid in pharmaceutical formulations and in spiked human urine on GCE. In all electrochemical methods, linezolid exhibited one well-defined oxidation peak in BR buffer at pH 7.

1.3.1 CYCLIC VOLTAMMETRIC BEHAVIOR

Cyclic voltammograms of linezolid were recorded within a wide range of the potential (0.4–1.2 V vs Ag/AgCl electrode) at different pH values, scan rates, and concentration. Linezolid gave a quasi-reversible oxidation peak. Figure 1.2 depicts cyclic voltammograms of 1.0×10^{-4} M linezolid in BR buffer (0.4 M pH 7) on GCE. The initial parameters of CV are as follows: initial potential (E_i) = 0.4 V, high potential (E_h) = 1.2 V, low potential (E_L) = 0 V, scan rate = 100 mV/s, quit time = 5 s, and sensitivity = 1×10^{-5} A/V.

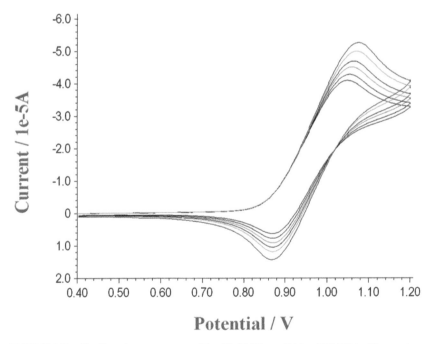

FIGURE 1.2 Cyclic voltammograms of 1×10^{-4} M linezolid in pH 7 BR buffer on glassy carbon electrode at different scan rates: (1) 50, (2) 100, (3) 150, (4) 200, (5) 250, and (6) 300 mV/s.

1.3.2   EFFECT OF SCAN RATE

The cyclic voltammograms of 1.0×10^{-4} M linezolid were recorded at different scan rates at 25°C (Fig. 1.2). Both peak potential (E_p) and peak current (i_p) were affected by scan rate v. The whole procedure for CV was repeated for linezolid with different scan rates ranging from 50 to 300 mV/s, while other parameters being kept constant. For a reversible system, the peak potential (E_p) is independent of the scan rate (v), but in our study we found that peak potential (E_p) shift slightly toward more positive potential as we increase the scan rate (v) that indicates that the process is quasi-reversible. For a reversible process, the peak current (i_p) is given by *Randles–Sevcik* equation.[34,35]

$$i_p = (2.69 \times 10^5) \, n^{3/2} \, A C_o D^{1/2} v^{1/2} \qquad (1.1)$$

where n is the number of electrons taking part in the rate determining step of electrode process, A is the apparent surface area of the electrode (in cm^2), C_o is the concentration of the electro active species (in mol/cm^3), D is the diffusion coefficient of the electro-active species (in cm^2/s), and v is the potential scan rate (V/s).

The effect of potential scan rate on the peak current of linezolid was studied, and peak current was proportional to the square root of scan rate ($v^{1/2}$) at the pH 7 (Fig. 1.3). Plot of the peak current versus square root of the scan rates ($v^{1/2}$) is described by the following equations:

$$i_{pa} = 0.438 \, v^{1/2} + 2.9115; \quad R^2 = 0.9917 \text{ [anodic region]} \qquad (1.2)$$
$$i_{pc} = 0.729 \, v^{1/2} + 1.4202; \quad R^2 = 0.999 \text{ [cathodic region]} \qquad (1.3)$$

As the scan rate was increased from 50 to 300 mV/s at a fixed concentration of linezolid, (1) the peak current increased steadily and (2) the peak current function ($i_p/A \, C \, v^{1/2}$) exhibited nearly constancy. The linear relationship existing between oxidation peak current (i_{pa}) and square root of scan rate ($v^{1/2}$) with slope 0.438 showed that the oxidation process is predominantly diffusion controlled in the whole scan rate range studied (Fig. 1.4). The variation of logarithm of peak current ($\log i_p$) versus logarithm of scan rate ($\log v$) can be expressed as follows:

$$\log i_p = 0.736 \log v + 0.3812; \quad R^2 = 0.9954 \text{ [anodic region]} \qquad (1.4)$$
$$\log i_p = 0.1823 \log v - 0.0279; \quad R^2 = 0.9935 \text{ [cathodic region]} \qquad (1.5)$$

All these certitude toward the diffusion-controlled nature of the electrode process.

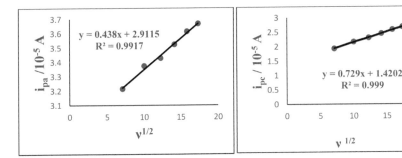

FIGURE 1.3   Plots of *anodic peak current* (i_{pa}) and *cathodic peak current* (i_{pc}) as a function of root of scan rates ($v^{1/2}$) for 1×10^{-4} M linezolid in pH 7 BR buffer.

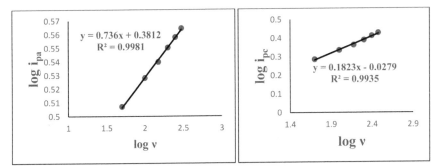

FIGURE 1.4   Plots of logarithm of peak current (log i_{pa} and log i_{pc}) as a function of logarithm of scan rate (log v) for 1×10^{-4} M linezolid in pH 7 BR buffer.

1.3.3   EFFECT OF CONCENTRATION

The cyclic voltammograms were recorded at different concentrations in pH 7 BR buffer (Fig. 1.5).

The peak current increased linearly with an increase in concentration of linezolid (Fig. 1.6). The *Randles–Sevcik* equation also indicates that peak current (i_p) is directly proportional to the concentration of linezolid C_o.[36,37]

$$i_p = (2.69 \times 10^5)\, n^{3/2}\, AC_o D^{1/2} v^{1/2} \tag{1.6}$$

The plot of peak current (i_p) versus concentration for linezolid yields a straight line that can be expressed by the following equations:

$$i_{pa} = 0.1357 \times 10^{-4}\,(\text{M}) + 3.6767; \quad R^2 = 0.9909 \text{ [anodic region]} \tag{1.7}$$

$$i_{pc} = 0.1373 \times 10^{-4}\,(\text{M}) + 1.6818; \quad R^2 = 0.9719 \text{ [cathodic region]} \tag{1.8}$$

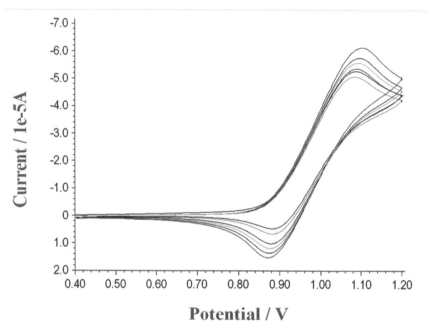

FIGURE 1.5   Cyclic voltammograms of linezolid in pH 7 BR buffer on glassy carbon electrode at different concentrations (1×10^{-4} to 6×10^{-4} M).

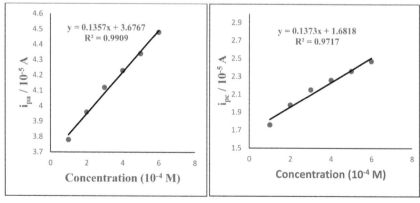

FIGURE 1.6   Plots of *anodic peak current* (i_{pa}) and *cathodic peak current* (i_{pc}) as a function of concentration of linezolid at scan rate of 100 mV/s.

The peak current and peak potential at different concentrations are shown in Table 1.1.

TABLE 1.1   Effect of Concentration of Linezolid on Cyclic Voltammetric Parameters in BR Buffer at pH 7.

S. No.	Concentration (M)	Anodic region		Cathodic region	
		E_p (V)	i_p (10^{-5} A)	E_p (V)	i_p (10^{-5} A)
1	1×10^{-4}	+1.081	3.78	−0.890	1.758
2	2×10^{-4}	+1.086	3.96	−0.891	1.979
3	3×10^{-4}	+1.089	4.12	−0.893	2.151
4	4×10^{-4}	+1.085	4.23	−0.891	2.258
5	5×10^{-4}	+1.083	4.34	−0.892	2.35
6	6×10^{-4}	+1.088	4.48	−0.891	2.47

1.3.4   ASSAY OF LINEZOLID IN PHARMACEUTICAL FORMULATION

Stripping voltammetric methods were optimized for trace determination of linezolid by differential pulse and square wave potential waveforms. Stripping voltammograms of bulk linezolid in BR buffer were recorded by SW–ASV and DP–ASV following its pre-concentration on the GCE by adsorptive accumulation for 15 s exhibiting a well-defined anodic peak with a better enhanced peak current magnitude at pH 7. The optimum operational parameters for the determination of bulk linezolid applying DP–ASV and SW–ASV were summarized in Table 1.2.

1.3.5   VALIDATION OF THE PROCEDURE

Validation of the proposed procedures for assay of the drug at trace levels was examined via the evaluation of the limit of detection (LOD), limit of quantization (LOQ), linearity, recovery, and repeatability.

1.3.6   LOD AND LOQ

The LOD and LOQ of linezolid were calculated using the following equations[38–40]:

$$LOD = 3s/b \tag{1.9}$$

$$LOQ = 10s/b \tag{1.10}$$

where s is the standard deviation of the intercept and b is the slope of the calibration curve.

TABLE 1.2   The Optimized Experimental Parameters of DP–ASV and SW–ASV Procedure.

DP–ASV		SW–ASV	
Parameter	Optimized value	Parameter	Optimized value
pH	7.0	pH	6.0
Buffer type	BR buffer	Buffer type	BR buffer
Strength of the buffer (M)	0.04	Strength of the buffer (M)	0.04
Temperature (°C)	22–25	Temperature (°C)	22–25
Initial potential (V)	0.4	Initial potential (V)	0.4
Final potential (V)	1.2	Final potential (V)	1.2
Scan increment (V)	0.004	Scan increment (V)	0.004
Pulse amplitude (V)	0.025	Pulse width (s)	0.2
Frequency (Hz)	15	Sample width (s)	0.0
Deposition time (s)	15	Pulse period(s)	0.2
		Amplitude (V)	0.05
		Quiet time (s)	2
		Deposition time (s)	15

An LOD of 2.199×10^{-7} M and an LOQ of 7.329×10^{-7} M bulk linezolid were achieved applying the described DP–ASV method. An LOD of 8.23×10^{-7} M and an LOQ of 2.74×10^{-6} M bulk linezolid were achieved applying the described SW–ASV method (Table 1.3).

TABLE 1.3   Stripping Voltammetric Determination of Linezolid in Bulk form Using DP–ASV and SW–ASV Modes.

Techniques	DP–ASV	SW–ASV
Linearity range (M)	1×10^{-7} to 5×10^{-7}	1×10^{-7} to 5×10^{-7}
Slope (A/M)	0.585	0.915
Intercept (10^{-5} A)	−0.0379	0.4388
Correlation coefficient	0.9852	0.9976
LOD (M)	2.119×10^{-7}	8.23×10^{-7}
LOQ (M)	7.329×10^{-7}	2.744×10^{-6}

1.3.7 LINEARITY

The applicability of the proposed DP–ASV and SW–ASV procedures as analytical methods for the determination of linezolid was examined by measuring the stripping peak current as a function of concentration of the bulk drug at least 3 times under the optimized operational parameters [Figs 1.7(A) and 1.8(A)]. The calibration plot of the peak current versus the concentration was found linear for the stripping voltammetric process [Figs 1.7(B) and 1.8(B)].

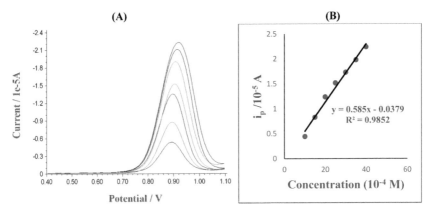

FIGURE 1.7   (A) DPAS voltammograms for linezolid and (B) plot of peak current (i_p) as a function of concentration of linezolid in pH 7 BR buffer.

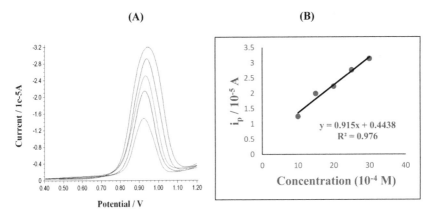

FIGURE 1.8   (A) SWAS voltammograms for linezolid and (B) plot of peak current (i_p) as a function of concentration of linezolid in pH 7 BR buffer.

The linear regression equation is expressed as follows:

DP–ASV: i_p $(10^{-5}\,A) = 0.585 \times 10^{-4}\,M + 0.0379;$ $R^2 = 0.9852$ (1.11)

SW–ASV: i_p $(10^{-5}\,A) = 0.915 \times 10^{-4}\,M + 0.4388;$ $R^2 = 0.9760$ (1.12)

The regression plots showed that there is a linear dependence of the current intensity on the concentration in both DP–ASV and SW–ASV modes over the range, as given in Table 1.3. The table also shows the detection limits and the results of the analysis of the experimental data such as slopes, intercept, and correlation coefficients.

1.3.8   REPEATABILITY

The repeatability was examined by performing five replicate measurements for 1×10^{-6} M bulk drug followed pre-concentration for 30 s under the same operational conditions. Percentage recoveries of 100.5, 99.25, 99.5, 99.63, and 98.6 were achieved with a mean value of 99.55 and RSD (%) of 0.577, which indicates repeatability and high precision for SW–ASV method. On the other hand, percentage recoveries of 97, 98.66, 99.4, 99.85, and 98.77 were achieved with a mean value of 98.73 and RSD (%) of 1.09, which indicates repeatability and high precision for DP–ASV method.

1.4   ASSAY OF LINEZOLID IN SPIKED HUMAN URINE

Linezolid in spiked human urine was successfully analyzed by both the previously described voltammetric methods (DP–ASV and SW–ASV). Representative DP–AS and SW–AS voltammograms of linezolid in spiked human urine were recorded under the optimum operational conditions of the described stripping voltammetric methods as shown in Figures 1.9(A) and 1.10(A), respectively.

The calibration plot of the peak current versus the concentration was found linear for the stripping voltammetric process [Figs. 1.9(B) and 1.10(B)].

The linear regression equation is expressed as follows:

$$\text{DP–ASV: } i_p\,(10^{-5}\,\text{A}) = 0.778 \times 10^{-5}\,\text{M} + 0.0161; \quad R^2 = 0.9906 \quad (1.13)$$

$$\text{SW–ASV: } i_p\,(10^{-5}\,\text{A}) = 2.759 \times 10^{-5}\,\text{M} + 1.2701; \quad R^2 = 0.9741 \quad (1.14)$$

Detection limits of 2.614×10^{-7} and 2.361×10^{-6} M and quantitation limits of 8.71×10^{-7} and 7.896×10^{-7} M linezolid were achieved by the described DP–ASV and SW–ASV methods (Table 1.4), respectively. Mean percentage recoveries and relative standard deviations of 99.54 ± 0.9413 (DP–ASV) and 98.98 ± 0.443 (SW–ASV) were achieved on the basis of replicate measurements of 1×10^{-5} M linezolid in spiked human urine. These results confirmed the reliability of the described stripping voltammetric methods for the assay of linezolid in spiked human urine.

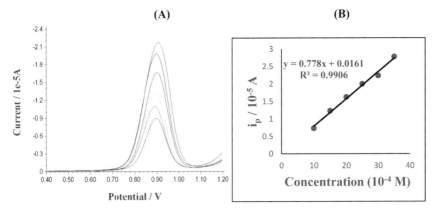

FIGURE 1.9 (A) DPAS voltammograms of linezolid in spiked human urine sample and (B) plot of peak current (i_p) as a function of concentration of linezolid in spiked human urine.

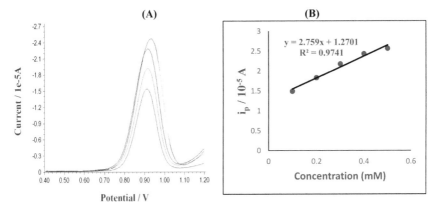

FIGURE 1.10 (A) SWAS Voltammograms of linezolid in spiked human urine samples and (B) plot of peak current (i_p) as a function of concentration of linezolid in spiked human urine.

TABLE 1.4 Stripping Voltammetric Determination of Linezolid in Spiked Human Urine Using DP–ASV and SW–ASV Modes.

Techniques	DP–ASV	SW–ASV
Linearity range (mol/L)	1×10^{-7} to 5×10^{-5}	1×10^{-7} to 5×10^{-5}
Slope (A/M)	0.778	2.759
Intercept (10^{-5} A)	0.0161	1.270
Correlation coefficient (R^2)	0.9906	0.9974
LOD (M)	2.614×10^{-7}	2.361×10^{-6}
LOQ (M)	8.713×10^{-7}	7.869×10^{-6}

1.5 CONCLUSION

A fully validated, sensitive, selective, fast, and low-cost anodic stripping square wave voltammetry and anodic stripping differential pulse voltammetry procedure were developed for determining the trace of linezolid in bulk form and in spiked human urine. The electro-oxidation of linezolid at the GCE in pH 7 BR buffer solution was studied and discussed without the necessity for extraction or formation of prior to analysis. The achieved LOQ and LOD by means of the described stripping voltammetric methods are low as well as they offer good possibilities for the determination of drug in low-dosage pharmaceutical preparations and in real plasma samples. The described methods could be recommended for use in quality control and clinical laboratories.

ACKNOWLEDGMENT

The authors gratefully acknowledge financial support by Council of Scientific and Industrial Research (09/149/0573).

KEYWORDS

- **linezolid**
- **antibacterial**
- **cyclic voltammetry (CV)**

- square wave anodic stripping voltammetry (SW–ASV)
- glassy carbon electrode (GCE)
- diffusion controlled
- Britton–Robinson (BR) buffer

REFERENCES

1. Linezolid Monograph for Professionals. *Drugs.com\Prescription Drug Information, Interactions & Side Effects* [Online]. https://www.drugs.com/monograph/linezolid. html (accessed July 07, 2017).
2. Shinabarger, D. L.; Marotti, K. R.; Murray, R. W. *Antimicrob. Agents Chemother.* **1997,** *10,* 2132–2136.
3. Brickner, S. J. *Curr. Pharm. Des.* **1996,** *2,* 175–194.
4. Aoki, H.; Ke, L.; Poppe, S. M.; Poel, T. J.; Weaver, E. A.; Gadwood, R. C. *Antimicrob. Agents Chemother.* **2005,** *46,* 1080–1085.
5. Munoz, J. L.; Gutierrez, M. N.; Sanchez, F. J.; Yague, G.; Segovia, M.; Garcia-Rodriguez, J. A. *Int. J. Antimicrob. Agents* **2002,** *20,* 61–64.
6. Wilson, A. P.; Cepeda, J. A.; Hayman, S.; Whitehouse, T.; Singer, M.; Bellingan, G. *J. Antimicrob. Chemother.* **2006,** *58,* 470–473.
7. Indian Pharmacopoeia. *Indian Pharmacopoeia Commission, Government of India*; Addendum: New Delhi, 2008; pp 2377–2378.
8. Patel, M. M.; Patel, D. P.; Goswami, K. *Int. J. Pharm. Res. Scholars* **2012,** *1,* 112–118.
9. Patel, S. A.; Patel, P. U.; Patel, N. J.; Patel, M. M.; Bangoriya, U. V. *J. AOAC Int.* **2007,** *90* (5), 1272–1277.
10. Serena, F.; Gennaro, D. P.; Enzo, R.; Massimo, A.; Pierluigi, N. *Biomed. Chromatogr.* **2013,** *27* (11), 1489–1496.
11. Jerome, T. *J. Chromatogr. B* **2004,** *813* (1–2), 145–150.
12. Lories, I. B. *Anal. Lett.* **2003,** *36* (6), 1147–1161.
13. Prasanti, K. J.; Sundar, B. S. *Int. J. Pharm. Biol. Sci.* **2012,** *3* (3), 44–51.
14. Patel, S. A.; Patel, P. U.; Patel, N. J.; Patel, M. M.; Bangoriya, U. V. *Indian J. Pharm. Sci.* **2007,** *69* (4), 571–574.
15. Ozkan, S. A. *Electroanalytical Methods in Pharmaceutical Analysis and Their Validation*; HNB Publishing: New York, NY, 2011.
16. Scholz, F. *Electroanalytical Methods: Guide to Experiments and Applications*; Springer: New York, NY, 2010.
17. Gupta, V. K.; Jain, R.; Radhapyari, K.; Jadon, N.; Agarwal, S. *Anal. Biochem.* **2011,** *408,* 179–196.
18. Bard, A. J.; Faulkner, L. R. *Electrochemical Methods: Fundamental and Application*; John Wiley and Sons: New York, NY, 2006.
19. Monk, P. M. S. *Fundamentals of Electroanalytical Chemistry*; John Wiley and Sons: New York, NY, 2001.

20. Bond, A. M. *Broadening Electrochemical Horizons: Principles and Illustration of Voltammetric and Related Techniques*; Oxford University Press: New York, NY, 2002.
21. Ozoemena, K. I. *Recent Advances in Analytical Electrochemistry*; Transworld Research Network: Trivandrum, 2007.
22. Dogan-Topal, B.; Ozkan, S. A.; Uslu, B. *Open Chem. Biomed. Methods J.* **2010**, *3*, 56–73.
23. Wang, J. *Analytical Electrochemistry*, 2nd ed.; Wiley-VCH: New York, NY, 2002.
24. Wang, J. *Stripping Analysis: Principles, Instrumentation and Applications*; VCH Inc.: Deerfield Beach, FL, 1985.
25. Patriarche, G. J.; Zhang, H. *Electroanalysis* **2005**, *8*, 573–579.
26. El-Maali, N. A. *Bioelectrochemistry* **2004**, *64* (1), 99–107.
27. Ozkan, S. A.; Uslu, B.; Aboul-Enein, H. Y. *Crit. Rev. Anal. Chem.* **2003**, *133* (3), 155–181.
28. Adams, R. N. *J. Pharm. Sci.* **2006**, *58* (10), 1171–1184.
29. Ozkan, S. A. *Curr. Pharm. Anal.* **2009**, *5*, 127–143.
30. Xu, Q.; Yuan, A.; Zhang, R.; Bian, X.; Chen, D.; Hu, X. *Curr. Pharm. Anal.* **2009**, *5*, 144–155.
31. Uslu, B.; Ozkan, S. A. *Comb. Chem. High Throughput Screening* **2007**, *10*, 495–513.
32. Wang, J. *Electroanalytical Chemistry*; Wiley-VCH Publication: New York, NY, 2006.
33. Bard, A. J.; Faulkner, L. R. *Electrochemical Methods: Fundamentals and Applications*; John Wiley and Sons Inc.: New York, NY, 2001.
34. Kissinger, P. T.; Heineman, W. R. *Laboratory Techniques in Electroanalytical Chemistry*; Marcel Dekker: New York, NY, 1996.
35. Wang, J. *Analytical Electrochemistry*, 2nd ed.; Wiley-VCH: New York, NY, 2002.
36. Wang, J. *Stripping Analysis: Principles, Instrumentation and Applications*; VCH Publishers Inc.: Deerfield Beach, FL, 1985.
37. Gosser, D. K. *Cyclic Voltammetry*; VCH: New York, NY, 1994.
38. Brett, C. M. A.; Brett, A. M. O. *Electrochemistry: Principles Methods & Applications*; Oxford University Press: Oxford, 1993.
39. Swartz, M. E.; Krull, I. S. *Analytical Method Development and Validation*; Marcel Dekker: New York, NY, 1997.
40. Riley, C. M.; Rosanske, T. W. *Development and Validation of Analytical Methods*; Elsevier Science Ltd.: New York, NY, 1996.

CHAPTER 2

Recent Highlights in Electrophilic Fluorination with 1-Fluoro-2,4,6-Trimethylpyridinium Tetrafluoroborate and NFSI (*N*-Fluorobenzenesulfonimide) Reagents Directed Toward the Synthesis of Heterocycles

PANKAJ SHARMA[1], NUTAN SHARMA[2], AMIT KUMAR[1], GUNJAN KASHYAP[1], and SUNITA BHAGAT[1*]

[1]*Department of Chemistry, ARSD College, University of Delhi, Dhaula Kuan, New Delhi 110021, India*

[2]*Department of Chemistry, Faculty of Science, Shree Guru Gobind Singh Tricentenary University, Gurugram-122505, Haryana, India*

[*]Corresponding author. E-mail: sunitabhagat28@gmail.com

ABSTRACT

In this chapter, we have summarized an overview of recent highlights in electrophilic fluorination with 1-fluoro-2,4,6-trimethylpyridinium tetra-fluoroborate and NFSI (*N*-fluorobenzenesulfonamide) reagents which are widely used in the synthesis of different heterocycles.

2.1 INTRODUCTION

The development of fluorination chemistry began more than 100 years ago with the first examples of nucleophilic and electrophilic fluorination reactions being reported in the second half of the 19th century.[1a,b,2]

Fluorinated molecules play a significant role in pharmaceutical/medicinal,[3a,b] agrochemical,[3c,d] and material sciences[3e] due to the unique properties of fluorine atom.[4] The introduction of fluorine atom can modulate the properties of biologically active compounds, since this may lead to changes in lipophilicity, chemical reactivity, solubility, conformity, and ability of hydrogen bonding.[5] As a result, 30–40% of agrochemicals and 20–25% of pharmaceuticals on the market are estimated to contain fluorine.[6] The efficacy of agrochemicals and pharmaceuticals is often enhanced due to the presence of a fluorine atom in the structure.[7] Frequently, these types of compounds can be synthesized from small moieties in which fluorine is located at a specific site, or often it is desirable to introduce the element directly in order to obtain the desired chemical composition.[8,9]

Electrophilic fluorination is one of the most direct methods for selective introduction of fluorine into organic compounds.[10] In this respect, various reagents for electrophilic fluorination have been developed till now as shown in Figure 2.1. Elemental fluorine itself is one of the most powerful reagents.[11] However, fluoroxcompounds,[12] such as CF_3OF, $CF_3C(O)OF$, $CsSO_4F$, and $CH_3C(O)OF$[13] some of which are generated in situ, are exciting reagents for the introduction of fluorine electrophilically into a wide variety of organic compounds.[14,15] Most electrophilic fluorinating reagents are ultimately derived from fluorine gas, the strongest elemental oxidant known, which is synthesized by electrolysis of potassium difluoride in hydrogen fluoride.[16] Electrophilic fluorination reactions with highly oxidizing fluorinating reagents[17] such as fluorine gas, hypofluorites, fluoroxysulfates, and perchloryl fluoride are challenging to perform due to the high reactivity of the reagents. Xenon difluoride was developed as a more stable electrophilic fluorination source, but its high oxidation potential still limits the tolerance of this reagent to functional groups.[18] The development of crystalline, benchtop-stable fluorinating reagents such as N-fluorobis(phenyl)sulfonimide (NFSI)[15] and related analogs,[19,20] N-fluoropyridinium salts,[21] and 1-chloromethyl-4-fluoro-1,4-diazoniabicyclo-[2.2.2]octane bis(tetrafluoroborate) (Selectfluor, F-TEDABF$_4$)[22] was crucial for the development of selective, functional group–tolerant fluorination methods.

Another class of electrophilic fluorinating reagents with the general structure $R_2N–F$ or $R_3N^+–F$ began to gain popularity. In comparison to the previous reagents, these compounds were stable, safer, milder, and less expensive to produce. Furthermore, some of these compounds proved to be as reactive as established reagents in some cases, which were also capable of a degree of selectivity that was previously unattainable.[23] Efforts by Umemoto et al. led to the first isolatable *N*-fluoropyridinium salts, which had good activity and were amenable to commercial production.[24]

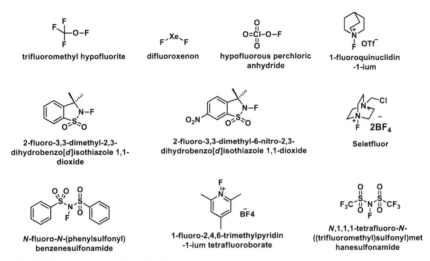

FIGURE 2.1   Electrophilic fluorinating reagents.

The important role of the counteranion, which influences the reactivity, selectivity, and stability of the reagent, was also demonstrated.[25] Des Marteau and coworkers later reported the discovery of *N*-fluorobis[(trifluoromethyl) sulfonyl]imide, accessed by the reaction of bis[(trifluoromethyl)sulfonyl] imide with fluorine gas.[26,27] The synthesis of this compound, which is still one of the most powerful sources of electrophilic fluorine, was a step forward in the identification of a stable source of electrophilic fluorine with desirable physical properties. One major drawback is that it is not commercially available, so that the use of fluorine gas for its preparation in the laboratory is ultimately required. More recently, the development of the reagent NFSI presented a major advance for electrophilic fluorination, as it is a reliable, mild, stable, and effective source of electrophilic fluorine that lends itself to large-scale synthesis and is commercially available. A number of reagents, each with its advantages and disadvantages, currently exist for the electrophilic incorporation of fluorine and have been the subject of several reviews.[28–31] In the present review, we give a comprehensive overview of the literature reports emphasizing two potential electrophilic fluorinating reagents 1-fluoro-2,4,6-trimethylpyridinium tetrafluoroborate and NFSI (*N*-fluorobenzenesulfonimide).

2.2   ELECTROPHILIC FLUORINATING REAGENTS

2.2.1   *N-FLUOROPYRIDINIUM TETRAFLUOROBORATE*

In Scheme 2.1, Lagow and coworkers used *N*-fluoropyridinium tetrafluoroborate as fluorinating reagent among the *N*-fluoropyridinium salt series and can react as an electrophilic fluorine source. Exhibiting almost the same reactivity as other related *N*-fluoropyridinium salts, albeit with a somewhat lower solubility, it has a high fluorine content (103 g kg^{-1}).[32]

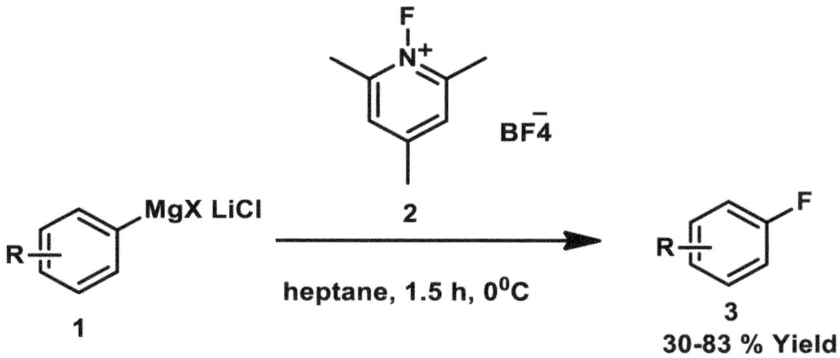

SCHEME 2.1   Synthesis of aryl fluoride from Grignard reagents.

Tomita and coworkers used *N*-fluoropyridinium tetrafluoroborate as fluorinating reagent as shown in Scheme 2.2. In this case of fluorination of 1-(trimethylsilyloxy) cyclohexene, the yields and/or rates of reaction are not satisfactory, compared to those of the triflate analog, because of the reduced solubility of fluorinating reagents in CH_2Cl_2. The same reactions performed in acetonitrile, however, give similar results[33] and efficient enantioselective electrophilic fluorination of β-ketoesters can be achieved. It can be seen that the use of NFPY (*N*-fluoropyridinium tetrafluoroborate) resulted in moderate enantioselectivities. Interestingly, the addition of the electron-poor alcohol HFIP (1,1,1,3,3,3-hexafluoroisopropanol) as additive led to an increase in enantioselectivity to 67% ee.[34]

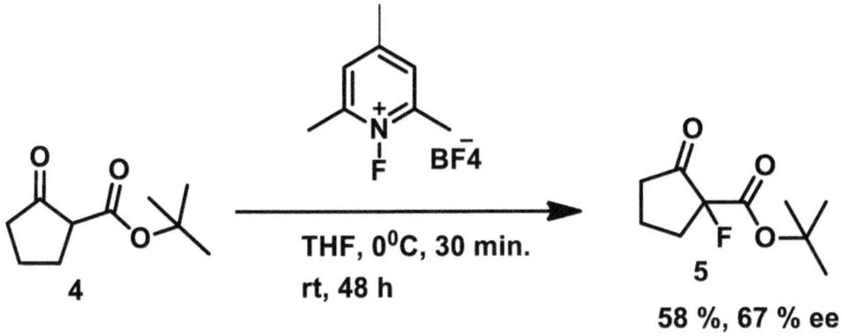

SCHEME 2.2 Fluorination using N-fluoropyridinium tetrafluoroborate.

1-Fluoro-2,4,6-trimethylpyridin-1-iumtetrafluoroborate was selected as the best electrophilic fluorine donor for palladium-catalyzed fluorination on 2-arylpyridines. In Scheme 2.3, Takeru Furuya and coworkers used palladium-catalyzed fluorination of organic molecules. The C–H activation/oxidative fluorination process allows *ortho*-fluorination of the aryl ring under microwave irradiation.[35]

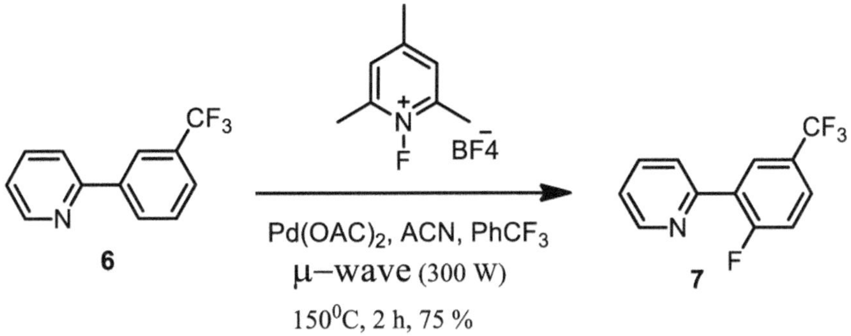

SCHEME 2.3 Palladium-catalyzed fluorination of carbon–hydrogen bonds.

In Scheme 2.4, Sanford and coworkers used palladium-catalyzed fluorination of 8-methylquinoline. Initial investigations focused on the Pd(OAc)$_2$-catalyzed benzylic fluorination of 8-methylquinoline. This substrate was selected because it undergoes facile quinoline-directed C–H activation at Pd(II) to generate a 6-benzyl-Pd species and, as such, has been shown to serve as an excellent substrate for related Pd-catalyzed C–H activation/oxidative functionalization reactions.[36,37] An initial screen of F^+ reagents under thermal reaction conditions (10 mol% of $Pd(OAc)_2$, 110 °C, 13 h, benzene) revealed that several conditions affected the desired benzylic C–H bond fluorination reaction, providing the desired compound in modest 9–36% yields while using different fluorinating reagents.[38] However, *N*-fluoro-2,4,6-trimethylpyridinium tetrafluoroborate served as a highly effective F^+ source affording products in a dramatically improved 82% isolated yield. Importantly, control reactions in the absence of Pd catalyst showed none of the fluorinated product.[39]

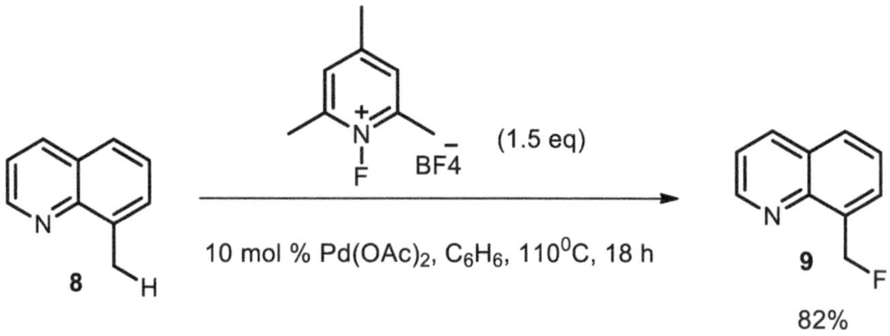

SCHEME 2.4   Palladium-catalyzed fluorination of 8-methylquinoline.

In Scheme 2.5, experimental investigation concludes a change of rate limiting step using 8-aminoquinoline versus 8-aminoquinoxaline as directing group by affecting the key Pd(II) to Pd(IV) oxidation step. This finding led to the discovery of a simple ligand capable of lowering the activation energy for the oxidation of Pd. This intriguing mechanistic journey provides new clues into oxidative C–H functionalization reactions. We expect these results will stimulate the development of new C–H functionalization reactions. It is being observed that if 8-aminoquinoline versus 8-aminoquinoxaline are removed from the reaction mixture, the rate of reaction is affected and the overall yield of the reaction affected by 42%. If the reaction is performed without ligand, the yield is only 15% and the yields vary with different ligands.[40]

Double fluorination through two subsequent *ortho*-fluorination events addressed with a weakly coordinating anionic *ortho*-directing group, *N*-perfluorotolylamide derived from benzoic acid that allows for rapid displacement of the monofluorinated product by the substrate, thus affording high selectivity for monofluorination (Scheme 2.6).

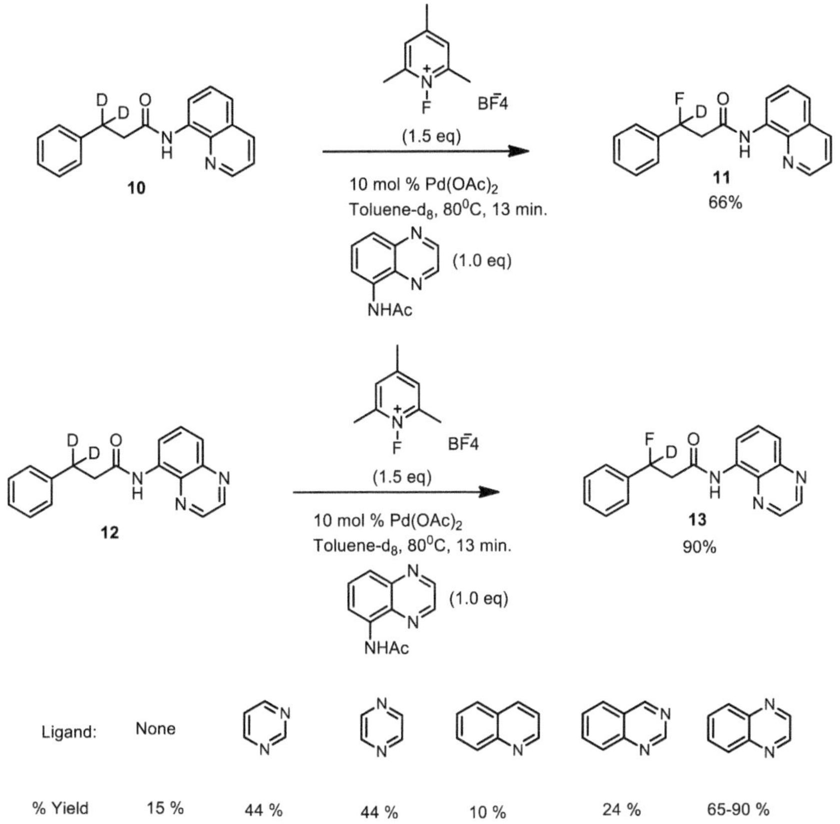

SCHEME 2.5   Ligand-assisted palladium(II)/(IV)o for sp³ C–H fluorination.

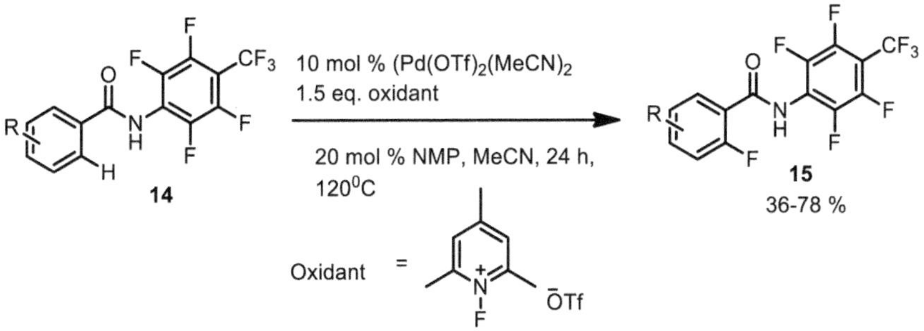

SCHEME 2.6   *N*-perfluorotolylamide-directed Pd-catalyzed fluorination of arenes.

2.2.2   ELECTROPHILIC FLUORINATION USING N-FLUORO-N-(PHENYLSULFONYL) BENZENE SULFONAMIDE

Mikiko Sodeoka and coworkers reported a highly efficient catalytic enantioselective fluorination of β-ketoesters (Scheme 2.7).[41] In the presence of a catalytic amount of chiral Pd complexes, a variety of β-ketoesters reacted with *N*-fluorobenzenesulfonimide (NFSI) in an alcoholic solvent such as EtOH to afford the corresponding fluorinated products with excellent enantioselectivity (up to 94% ee). Various substrates, including cyclic and acyclic β-ketophosphonates, underwent the reaction with *N*-fluorobenzenesulfonimide (NFSI) in EtOH to give the corresponding fluorinated products in a highly enantioselective manner (94–98% ee).

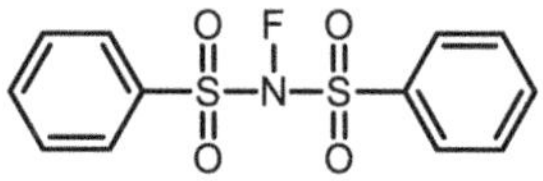

16
NFSI
N-fluoro-*N*-(phenylsulfonyl)benzenesulfonamide

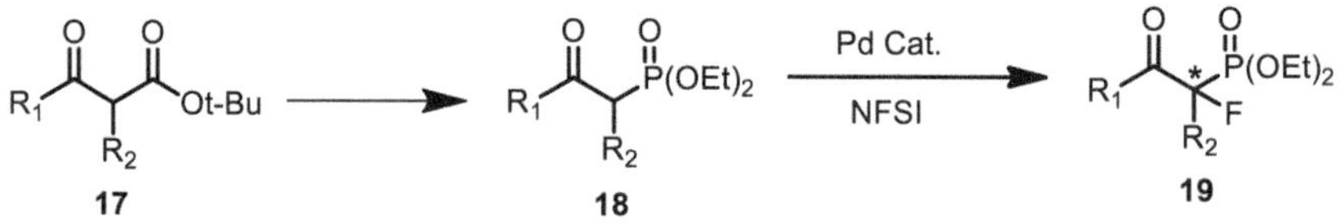

SCHEME 2.7   Enantioselective fluorination of β-ketoesters.

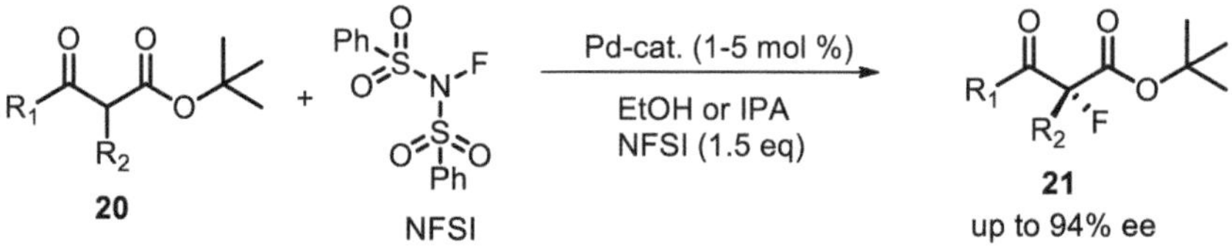

SCHEME 2.8   Catalytic enantioselective fluorination of β-ketoesters.

Dominique Cahard and coworkers developed a new efficient catalytic enantioselective electrophilic fluorination of acyclic (Scheme 8) and cyclic (Scheme 9) β-ketoesters.[42] As low as 1 mol% of chiral bis(oxazoline)–copper triflate complexes catalyze the fluorination by means of *N*-fluorobenzenesulfonimide (NFSI). The use of 1,1,1,3,3,3-hexafluoroisopropanol

(HFIP) is crucial for achieving high enantioselectivity.[43] Herein, a new efficient catalytic enantioselective electrophilic fluorination of both cyclic and acyclic β-ketoesters by means of chiral bis(oxazoline)–copper complexes leads to enantioenriched fluorinated compounds in high yields and good enantioselectivities (Scheme 2.10). The positive impact on the enantiomeric excess of achiral additives such as HFIP is also reported.[44,45]

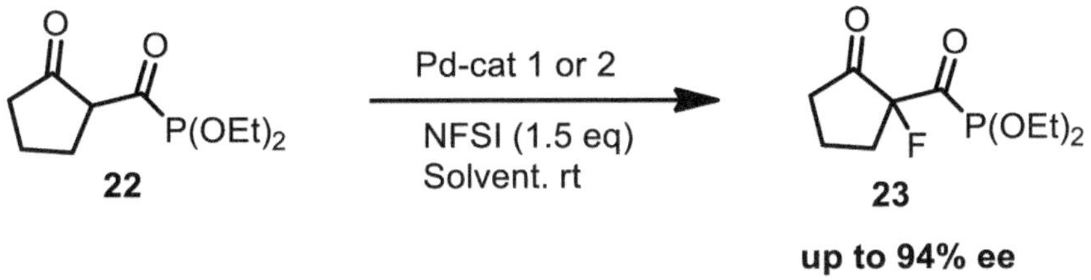

22 → Pd-cat 1 or 2 / NFSI (1.5 eq) / Solvent. rt → **23** up to 94% ee

SCHEME 2.9 Enantioselective electrophilic fluorination of cyclic β-ketoesters.

22 → chiral bis (oxazoline)-copper complexes / NFSI (1.5 eq) / Solvent. rt → **23**

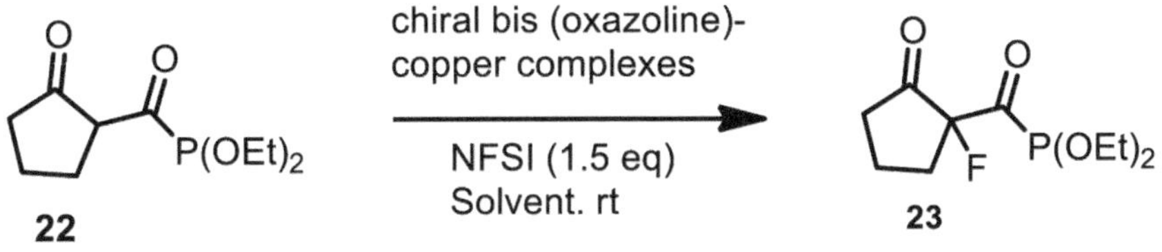

SCHEME 2.10 Copper-catalyzed enantioselective electrophilic fluorination of cyclic β-ketoesters.

In Scheme 2.11, fluorination of Grignard reagents with electrophilic *N*-fluorinated reagents NFSI is the most reliable method with simple aryl nucleophiles but is narrow in scope due to the basicity and nucleophilicity of the arylmagnesium reagents.[46,47] Through appropriate choice of solvent and reagents, undesired protodemetallation products can be minimized.[48] More recent work by Meng and Li demonstrated the regioselective *para*-fluorination of anilides with PhI(OPiv)$_2$ and hydrogen fluoride/pyridine.[49]

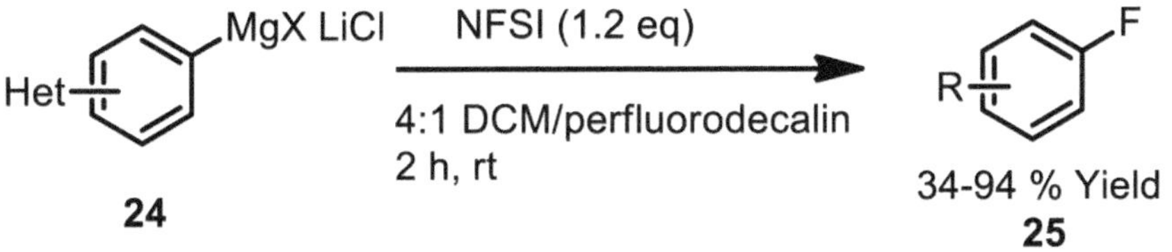

24 → MgX LiCl NFSI (1.2 eq) / 4:1 DCM/perfluorodecalin / 2 h, rt → **25** 34-94 % Yield

SCHEME 2.11 Synthesis of aryl fluorides from aryl Grignard reagents.

Several methods have exploited the two-point binding of dicarbonyl compounds to chiral Lewis acid complexes to control enantioselective fluorination. Asymmetric fluorination of β-ketoesters was achieved with Cu(II) (Scheme 2.12).[50]

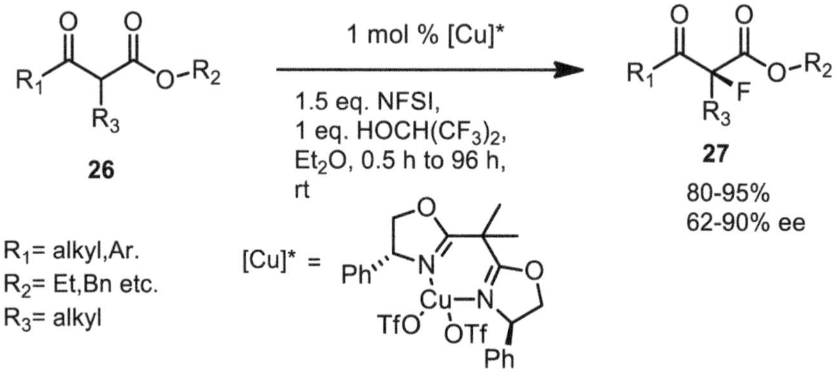

SCHEME 2.12   Cu-catalyzed asymmetric α-fluorination of β-ketoesters.

Several methods have exploited the two-point binding of dicarbonyl compounds to chiral Lewis acid complexes to control enantioselective fluorination. Asymmetric fluorination of β-ketoesters was achieved with Ni(II)-Box complexes (Scheme 2.13)[51] by Cahard and Shibata/Toru, respectively.

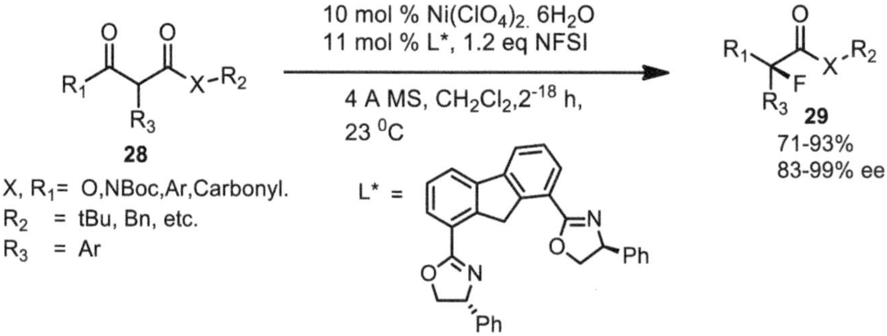

SCHEME 2.13   Ni-catalyzed asymmetric α-fluorination of β-ketoesters and *N*-Boc oxindoles.

Chiral bis(imino)bis(phosphine)ruthenium(II) complex by Togni (Scheme 2.14)[52] and scandium binaphthyl phosphate complexes by Inanaga.[53] The Ni-catalyzed reaction, using a 10 mol% catalyst loading, has demonstrated the broadest substrate scope so far and allows for α-fluorination of a variety of β-ketoesters and *N*-Boc-protected amides[51] in 71–93% yield and 83–99% ee, respectively. The catalytic enantioselective α-fluorination of α-substituted methyl, *tert*-butyl malonate was accomplished via chiral Lewis acid catalysis with Zn(II) acetate, (*R,R*)-4,6-dibenzofurandiyl-2,2¢-bis(4-phenyloxazoline) ligand, and NFSI. This approach was specifically optimized for the malonate substrate *en* route to the enantioselective synthesis of fluorinated β-lactams.[54,55]

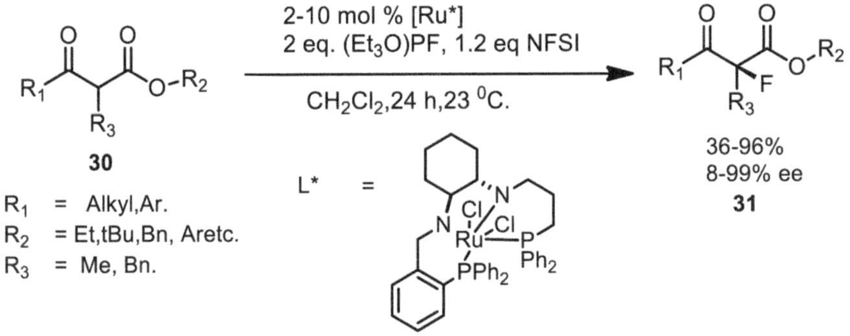

SCHEME 2.14   Ru-catalyzed asymmetric α-fluorination of β-ketoesters.

Chiral Pd–BINAP complexes developed by Sodeoka shown in Schemes 2.15 and 2.16 catalyze the enantioselective fluorination of α-ketoesters,[56] β-ketoesters (Scheme 2.11),[57] β-ketophosphonates,[58] oxindoles,[59] and α-ester lactones/lactams.[60] The use of chiral palladium complexes was particularly successful for the α-fluorination of acyclic α-ketoesters, cyclic and acyclic *tert*-butyl β-ketoester as well as oxindoles α-substituted with an electronically diverse range of aryl and alkyl groups.[56,57,59]

The aldehyde α-fluorination method described by MacMillan demonstrates a broader substrate scope, (Scheme 2.17).[61] These fluorinated compounds are useful for the synthesis of different biological active heterocycles.[62]

Jørgensen and coworkers used NFSI and organocatalytic α-fluorination of aldehydes. (Scheme 2.18)[63] while the method described the utilization of lower loadings of catalyst and electrophilic fluorinating reagent. Branched aldehydes constitute difficult substrates for enantioselective α-fluorination; nonetheless.

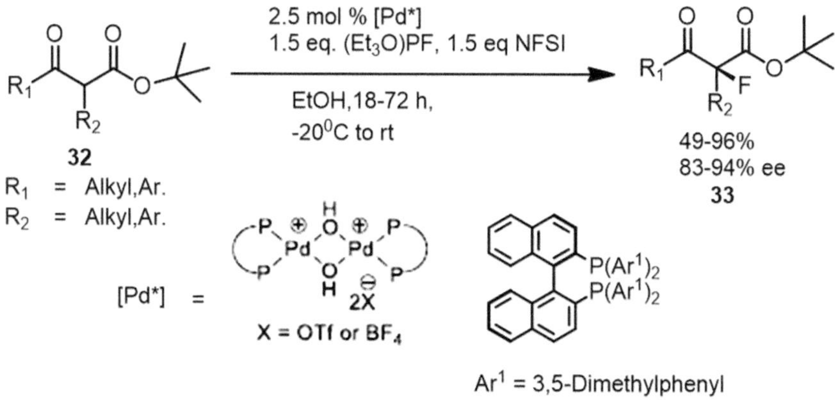

SCHEME 2.15   Pd-catalyzed asymmetric α-fluorination of β-ketoesters.

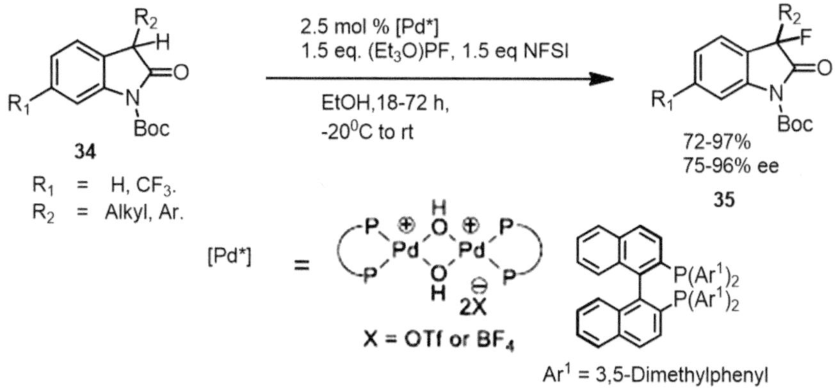

SCHEME 2.16   Pd-catalyzed asymmetric α-fluorination of oxindoles.

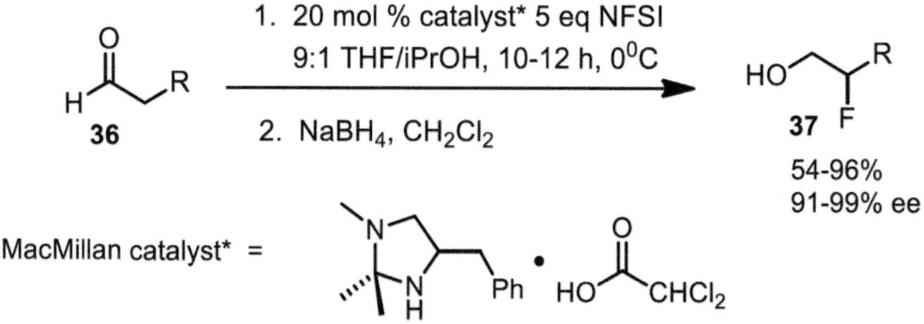

SCHEME 2.17   Organocatalytic asymmetric α-fluorination of aldehydes.

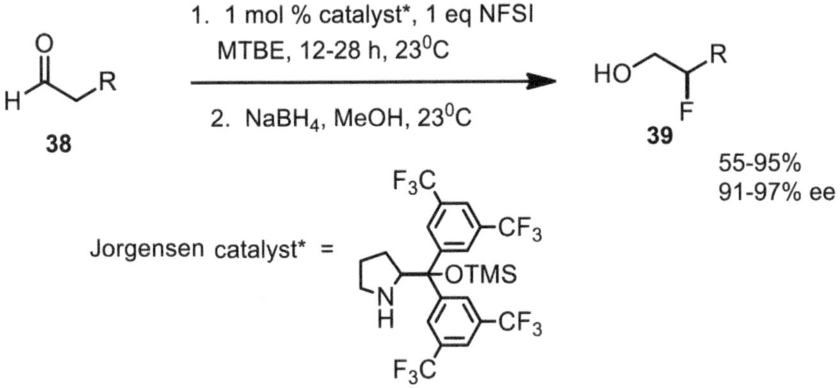

SCHEME 2.18   Organocatalytic asymmetric α-fluorination of aldehydes.

Barbas (Scheme 2.19)[64] accomplished the enantioselective organocatalytic α-fluorination of aldehydes. Similarly, enantioselective α-fluorination of ketones is possible using enamine catalysis.[65] Barbas reported a promising 98–99% yield and 45–66% ee for this class of substrates. The fluorinated aldehyde products are especially useful for the synthesis of enantiopure β-fluoroamines, which can be obtained by a chiral sulfinyl imine condensation, directed reduction sequence of the enantioenriched fluorinated aldehyde.[66]

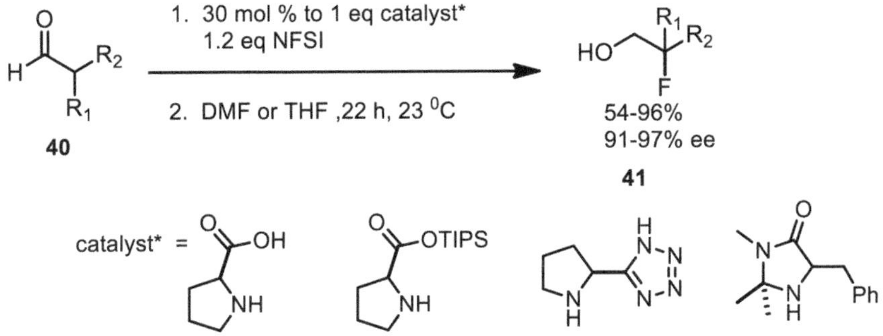

SCHEME 2.19   Organocatalytic asymmetric α-fluorination of α-branched aldehydes.

The N–F bonds in electrophilic fluorinating reagents have relatively low bond dissociation energies (2.84 eV for *N*-fluorosultam).[67] Under either photolysis or thermolysis conditions, a variety of *tert*-butyl alkyl peroxides afforded the corresponding alkyl fluorides upon treatment with NFSI (Scheme 2.20).[68] Primary alkyl fluoride formation was not efficient, which supports the mechanism hypothesis of radical intermediates.

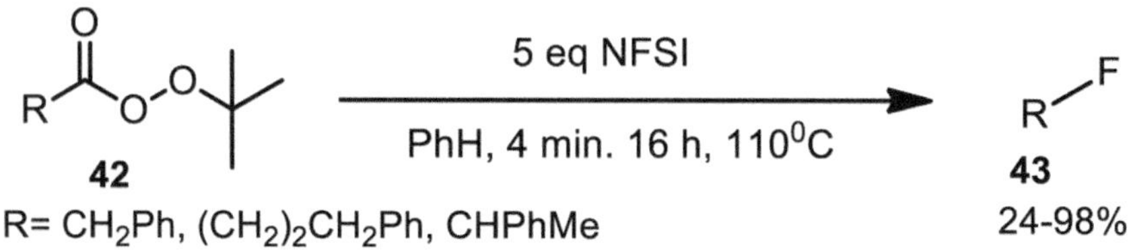

SCHEME 2.20   Fluorination of *tert*-butyl alkylperoxoates with NFSI.

The intermolecular aminofluorination of styrenes (Scheme 2.21)[69] can be facilitated by the use of palladium catalysts as described by Liu. Although reactions accomplish aminofluorination of alkenes, different approaches were employed and different reaction mechanisms proposed for each case. Intermolecular aminofluorination is thought to occur via fluoropalladation involving substrate, NFSI, and the active palladium complex, followed by oxidation to a putative Pd(IV) species, and subsequent reductive elimination to form a carbon–nitrogen bond. The formation of C–F bonds at sp^3 carbon centers by reductive elimination from high-valent transition metal complexes has also been investigated with Pt(IV) complexes.[70,71]

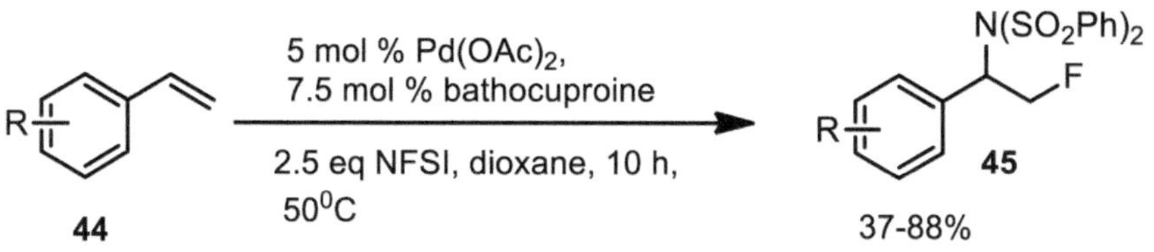

SCHEME 2.21   Intermolecular aminofluorination of styrenes with NFSI.

Enantiopure (*R*)-silyl ketones were prepared by diastereoselective silylation of the (*S*)- or (*R*)-1-amino-2-methoxy methylpyrrolidine (SAMP/RAMP) hydrazone and used as substrates in diastereoselective electrophilic fluorinations in which the silyl group acts as a traceless directing group. Lithium enolates of chiral compound generated by lithium diisopropyl amide (LDA) were fluorinated with enantiopure (*R*)-silyl ketones, were prepared by diastereoselective silylation of the (*S*)- or

(*R*)-1-amino-2-methoxymethylpyrrolidine (SAMP/RAMP) hydrazone, and used as substrates in diastereoselective electrophilic fluorinations in which the silyl group acts as a traceless directing group.[72] Lithium enolates of compound generated by LDA were fluorinated with NFSI in good yields and with high diastereomeric excesses. Interestingly, LiHMDS allowed reverse of diastereoselectivity to be obtained. The diastereoselectivity was found to reflect the ratio of enolate stereoisomer, with NFSI reacting only from the less sterically hindered enolate face. This concept was also applied to silyl enol ether; however, the fluorination gave rise to a significant amount of regioisomers (Scheme 2.22).[73]

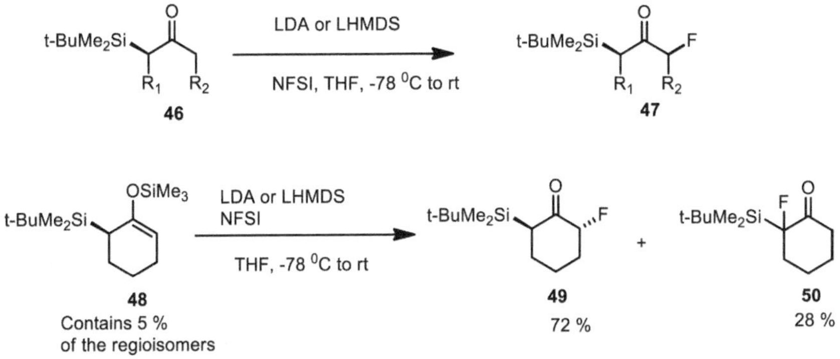

SCHEME 2.22   Electrophilic fluorination of (*R*)-silyl ketones.

A chiral oxazolidinone auxiliary was also used by Stauton and coworkers to direct the addition of a fluorine atom in the preparation of fluoro analog as a biosynthetic precursor of the ionophore antibiotic tetronasin (Scheme 2.23).[74]

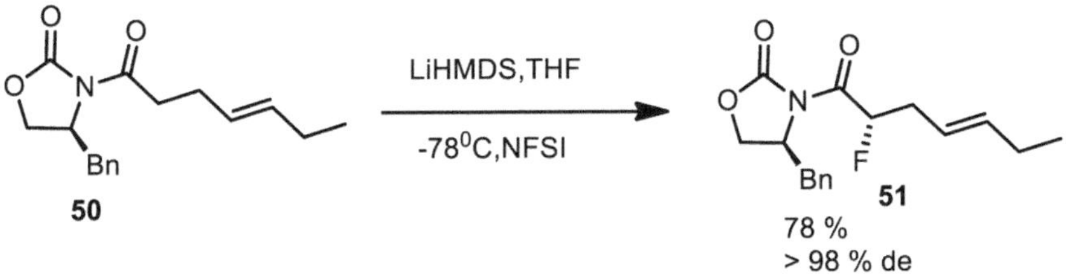

SCHEME 2.23   Fluorination of oxazolidinone.

A new catalytic stereoselective tandem transformation via Nazarov cyclization/electrophilic fluorination has been accomplished (Scheme 2.24). This sequence is efficiently catalyzed by a Cu(II) complex to afford fluorine-containing 1-indanone derivatives with two new stereocenters with high diastereoselectivity (*trans/cis* up to 49/1). Three examples of catalytic enantioselective tandem transformation are presented.[75]

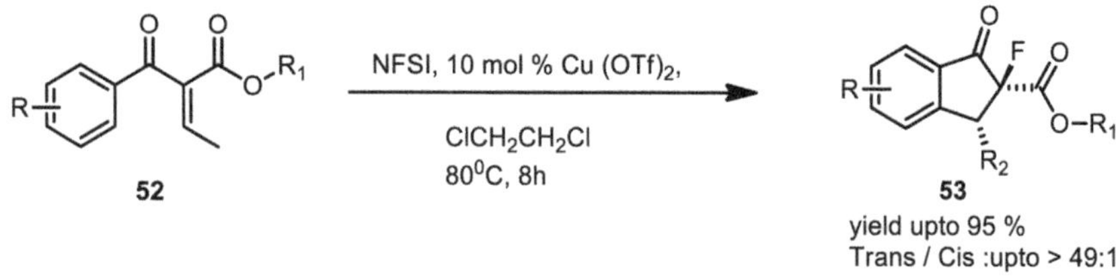

52 → NFSI, 10 mol % Cu (OTf)$_2$, ClCH$_2$CH$_2$Cl 80^{0}C, 8h → **53**

yield upto 95 %
Trans / Cis :upto > 49:1

SCHEME 2.24 Catalytic tandem Nazarov cyclization/sequential fluorination trapping.

In Scheme 2.25, Paul Knochel and coworkers developed a simple, convenient, and highly versatile one-pot method for converting aromatic and heteroaromatic bromides or iodides into the corresponding fluorides by choosing an optimized different solvent mixture. This procedure allows a direct access to fluorinated pyridines, thiophenes, pyrroles, and isoquinolines as well as to sterically congested fluorine-substituted benzenes, which are otherwise difficult to prepare. Further investigations of this potentially practical synthetic method are currently underway.[76]

54 → iprMgCl•LiCl THF,0^{0}C, 1h → **55** → NFSI (1.2 eq) Solvent. rt, 2h → **56**

SCHEME 2.25 Synthesis of aryl fluorides from aryl Grignard reagents.

The first organomediated asymmetric ^{18}F fluorination has been accomplished using a chiral imidazolidinone and [^{18}F] *N*-fluorobenzenesulfonimide in Scheme 2.26. This method provides access to enantioenriched ^{18}F-labeled α-fluoroaldehydes (>90% ee), which are versatile chiral ^{18}F synthons for the synthesis of radiotracers. The utility of this process is demonstrated with the synthesis of the PET (positron emission tomography) tracer (2*S*,4*S*)-4-[^{18}F] fluoroglutamic acid.[77]

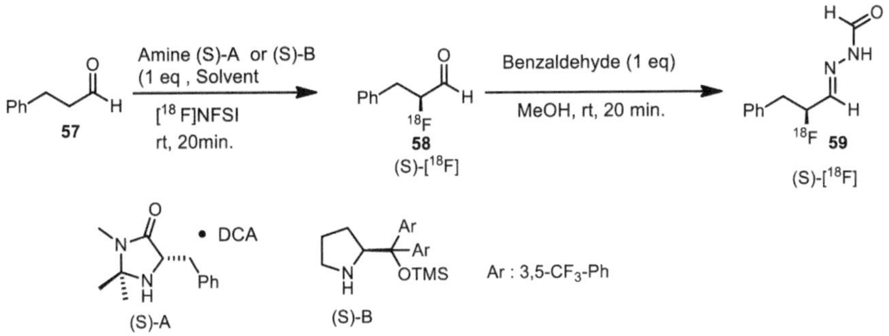

SCHEME 2.26 Organomediated enantioselective ^{18}F fluorination for PET applications.

Diastereoselective fluorination of α,β-unsaturated chiral oxazolidinone was conducted by reaction of LiHMDS followed by the addition of NFSI to produce a single diastereomer in 76% yield (Scheme 2.27). The complete diastereoselectivity reached with NFSI, compared to 82% de with NFOBS, was attributed to the greater steric bulk of NFSI. The reaction provided a nice example of deconjugative electrophilic fluorination. The acyclic fluoro compound was employed in the synthesis of fluoro carbohydrates.[78,79]

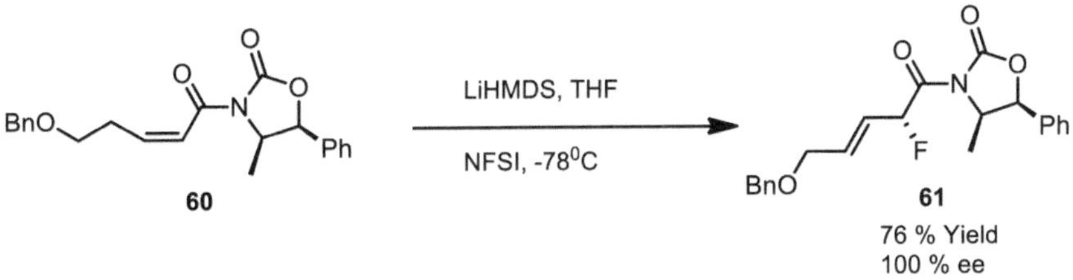

SCHEME 2.27 Diastereoselective fluorination of α,β-unsaturated chiral oxazolidinone.

Baoming and coworkers reports in Scheme 2.28 that a convenient and efficient method for fluorination of methylene cyclopropanes is reported. This is exemplified in the stereoselective preparation of *N*-[(*E*)-3-fluorobut-3-en-1-yl]-benzene sulfonamides by the reaction of methylene cyclopropanes with *N*-fluorobenzene sulfonimide (NFSI) in good to excellent yields.[80]

The allylic moieties have been found in many bioactive compounds and medicines, and compounds with a fluorine atom at the α-position of an allylic moiety, allylic fluorides, have exhibited excellent enhancement of the bioactivity of their parent compounds.[81,82] Furthermore, allyl fluorides have served as versatile intermediates in the synthesis of a large number of fluorinated compounds,[83,84] motivating the development of numerous synthetic methods with controlled regio- and stereoselectivity for allylic fluorides in Scheme 2.29.[85,86]

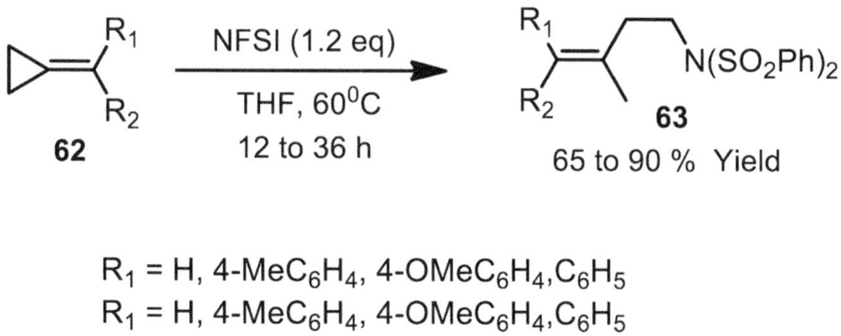
R1
R2
62
NFSI (1.2 eq)
THF, 60⁰C
12 to 36 h
R1
R2
N(SO2Ph)2
63
65 to 90 % Yield
R1 = H, 4-MeC6H4, 4-OMeC6H4,C6H5
R1 = H, 4-MeC6H4, 4-OMeC6H4,C6H5

SCHEME 2.28 Stereoselective fluorination of methylene cyclopropanes using NFSI.

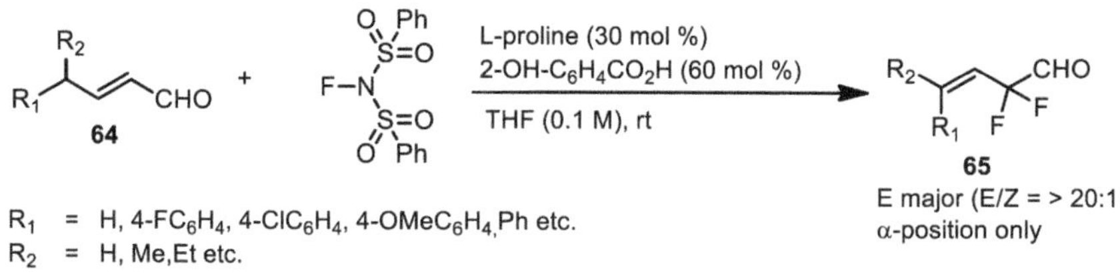
R2
R1
CHO
64
+
F–N
L-proline (30 mol %)
2-OH-C6H4CO2H (60 mol %)
THF (0.1 M), rt
R2
CHO
F
R1 F
65
E major (E/Z = > 20:1
α-position only
R1 = H, 4-FC6H4, 4-ClC6H4, 4-OMeC6H4,Ph etc.
R2 = H, Me,Et etc.

SCHEME 2.29 Stereoselective α,α-difluoro of γ,γ-disubstituted butenals.

The first catalytic α-fluoro allenoate synthesis is described by the complete structure with a suitable combination of *N*-heterocyclic carbine precatalyst, base, and fluorine reagent, the reaction proceeded smoothly to yield a wide range of α-fluoro allenoates with excellent chemoselectivity (Scheme 2.30). These substituted α-fluorinated allenoates have been synthesized for the first time, and they are versatile synthetic intermediates toward other useful fluorine-containing building blocks.[87]

The asymmetric fluorination of azolium enolates that are generated from readily available simple aliphatic aldehydes or α-chloro aldehydes and *N*-heterocyclic carbenes (NHCs) is described in Scheme 2.31. The process significantly expands the synthetic utility of NHC-catalyzed fluorination and provides facile access to a wide range of α-fluoro esters, amides, and thioesters with excellent enantioselectivity.[88] Catalytic systems that are based on *N*-heterocyclic carbenes (NHCs) have proven to be versatile in particular for establishing the α-stereogenic center of esters via key azolium enolate intermediates.[89]

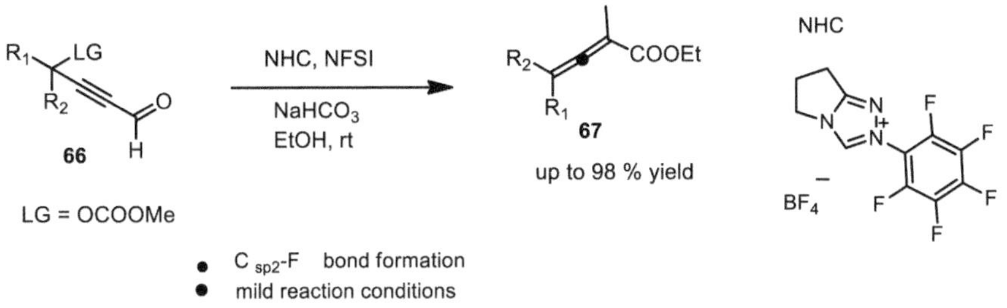

SCHEME 2.30   Catalytic α-fluoro allenoate synthesis.

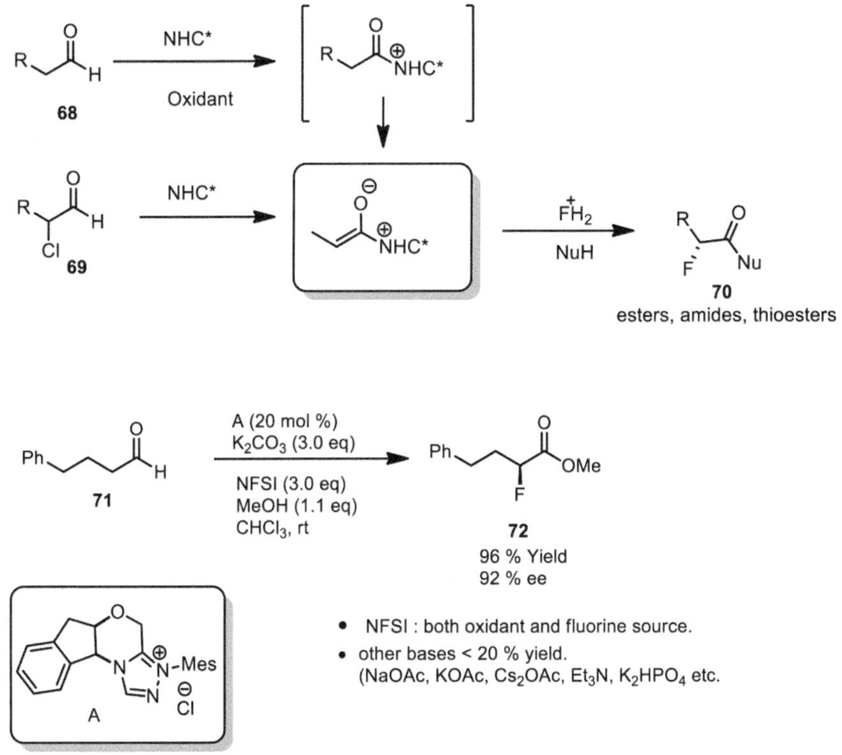

SCHEME 2.31   Fluorination of azolium enolates generated from simple aliphatic aldehydes and α-chloro aldehydes.

Guosheng Liu and coworkers reported a novel silver-catalyzed intramolecular amino fluorination of allenes for the synthesis of 4-fluoro-2,5-dihydropyrroles, in which the vinyl C–Ag bond is cleaved using NFSI to afford vinyl C–F bonds.[90] In addition, further convenient aromatization of fluorinated dihydropyrroles readily afforded the corresponding 4-fluoropyrrole derivatives (Scheme 2.32).

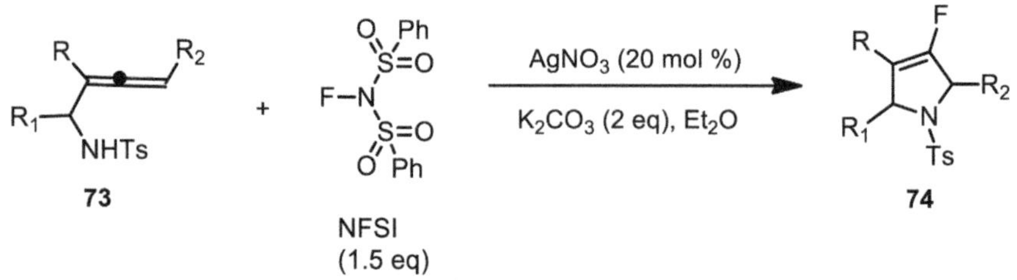

SCHEME 2.32   Silver-catalyzed intramolecular amino fluorination of allenes.

In Scheme 2.33, Prochiral indole **75** was also subjected to the optimized fluorocyclization conditions, which afforded the difluorinated tricyclic tetrahydro oxazolo[3,2-*a*] indole **76** in 50% yield, 60% ee.[91]

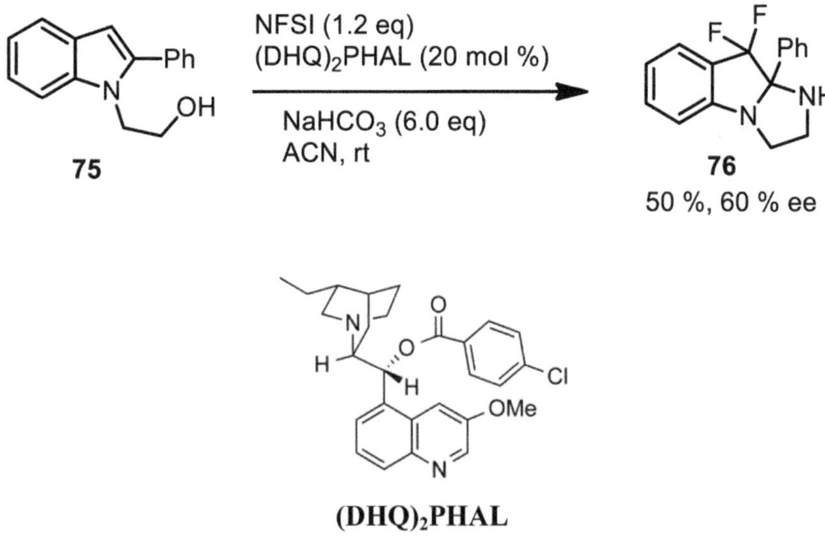

(DHQ)₂PHAL

SCHEME 2.33   Enantioselective fluorocyclization of prochiral indoles.

In Scheme 2.34, Jørgensen and coworkers reported a simple, direct one-pot organocatalytic approach to the formation of optically active propargylic fluorides **79**. This consists of organocatalytic α-fluorination of aldehydes followed by homologation with the Ohira–Bestmann reagent **78**, providing optically active propargylic fluorides.[92]

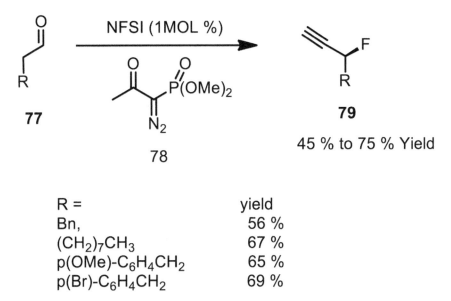

R =	yield
Bn,	56 %
$(CH_2)_7CH_3$	67 %
p(OMe)-$C_6H_4CH_2$	65 %
p(Br)-$C_6H_4CH_2$	69 %

SCHEME 2.34   One-pot synthesis of optically active propargylic fluorides.

2.3   CONCLUSION

The most significant, conceptual advances over the past decade in the area of fluorination, broadly defined, were made in the reactions that led to the formation of C–F bond, most prominently by organo and transition metal catalysis. The most challenging transformation remains the formation of the parent C–F bond, primarily due to the high hydration energy of fluoride, strong metal–fluorine bonds, and the highly polarized nature of bonds to fluorine. Reagents of the NF class, several of which are now commercially available, provide the organic chemist with a relatively safe and practical means of selectively positioning fluorine at chosen carbanionic-type sites in molecules. However, the stability of reagents in storage and ease of use are achieved at the cost of employing an R_2N or R_3N^+-organic carrier. For many large-scale uses, elemental fluorine, somehow "tamed" to act as a predictably selective electrophile, would ultimately be the most economical and environmentally "greener" alternative.

KEYWORDS

- **NFSI**
- **1-fluoro-2,4,6-trimethylpyridinium tetrafluoroborate**
- **electrophilic fluorination**

REFERENCES

1. (a) Swarts, F.; Bull, A. R. *Med. Belg.* **1898,** *35,* 375; (b) Borodine, A. *Ann. Chem. Pharm.* **1863,** *126,* 58.
2. Bchm, H. J.; Banner, D.; Bendels, S.; Kansy, M.; Kuhn, B.; Muller, K.; Obst-Sander, U.; Stahl, M. *ChemBioChem* **2004,** *5,* 637.
3. (a) Begue, J. P.; Delphon, D. B. *J. Fluorine Chem.* **2006,** *127,* 992; (b) Kirk, K. L. *J. Fluorine Chem.* **2006,** *127,* 1013; (c) Muller, K.; Faeh, C.; Diederich, F. *Science* **2007,** *317,* 1881; (d) Hagmann, W. K. *J. Med. Chem.* **2008,** *51,* 4359; (e) Purser, S.; Moore, P. R.; Swallow, S.; Gouverneur, V. *Chem. Soc. Rev.* **2008,** *37,* 320.
4. Review: Jeschke, P. *ChemBioChem* **2004,** *5,* 570.
5. Review: Pagliaro, M.; Ciriminna, R. *J. Mater. Chem.* **2005,** *15,* 4981.
6. Flynn, G. L. *Substituent Constants for Correlation Analysis in Chemistry and Biology*; John Wiley: New York, 1979.
7. Wilkinson, J. A. Recent Advances in the Selective Formation of the Carbon-Fluorine Bond. *Chem. Rev.* **1992,** *92,* 505.
8. Kirk, K. L. *J. Fluorine Chem.* **2006,** *127,* 1013.
9. German, L.; Zemskov, S. *Fluorinating Agents in Organic Synthesis*; Springer: Berlin, 1989.
10. Rozen, S. Elemental Fluorine As a Legitimate Reagent for Selective Fluorination of Organic Compounds. *Acc. Chem. Res.* **1988,** *21,* 307.
11. Rozen, S. Elemental Fluorine: Not Only for Fluoroorganic Chemistry! *Acc. Chem. Res.* **1996,** *29,* 243.
12. Taylor, S. D.; Kotoris, C. C.; Hum, G. Recent Advances in Electrophilic Fluorination. *Tetrahedron* **1999,** *55,* 12431.
13. Bailey, W. H.; Asteel, W. J.; Syvret, R. G. *Chem. Commun.* **2002,** *67,* 1416.
14. Patrick, T. B. *A Critical Review*; Hudlicky, M., Pavlath, A. E., Eds.; ACS Monograph 187; American Chemical Society: Washington, DC, 1995; p 133.

15. Rozen, S. *Chem. Rev.* **1996**, *96*, 1717.
16. Villalba, G.; Ayres, R. U.; Schroder, H. *J. Ind. Ecol.* **2007**, *11*, 85.
17. Sheppard, W. A. *Tetrahedron Lett.* **1969**, *10*, 83.
18. Tius, M. A. *Tetrahedron* **1995**, *51*, 6605.
19. Differding, E.; Ofner, H. *Synlett* **1991**, 187.
20. Barnette, W. E. *J. Am. Chem. Soc.* **1984**, *106*, 452.
21. Umemoto, T.; Kawada, K.; Tomita, K. *Tetrahedron Lett.* **1986**, *27*, 4465.
22. Banks, R. E.; Mohialdinkhaffaf, S. N.; Lal, G. S.; Sharif, I.; Syvret, R. G. *J. Chem. Soc. Chem. Commun.* **1992**, 595.
23. Paul, T. N.; Durn, S. G.; Burkart, M. D.; Vincent, S. P.; Wong, C. H. *Angew. Chem. Int. Ed.* **2005**, *44*, 192.
24. Umemoto, T.; Kawada, K.; Tomita, K. *Tetrahedron Lett.* **1986**, *27*, 4465.
25. Umemoto, T.; Fukami, S.; Tomizawa, G.; Harasawa, K.; Kawada, K., Tomita, K. *J. Am. Chem. Soc.* **1990**, *112*, 8563.
26. Resnati, G.; Des Marteau, D. D. *J. Org. Chem.* **1991**, *56*, 4925.
27. Des Marteau, D. D.; Xu, Z. Q.; Witz, M. *J. Org. Chem.* **1992**, *57*, 629.
28. Zupan, M.; Stavber, S. *Trends Org. Chem.* **1995**, *57*, 629.
29. Banks, R. E. *J. Fluorine Chem.* **1998**, *87*, 1.
30. Banks, R. E.; Besheesh, M. K.; Khaffaf, S. N. M.; Sharif, I. *J. Chem. Soc. Perkin Trans.* **1996**, *1*, 2069.
31. Lal, G. S.; Pez, G. P.; Syvret, R. G.; *Chem. Rev.* **1996**, *96*, 1737.
32. Anbarasan, P.; Neumann, H.; Beller, M. *Angew. Chem. Int. Ed.* **2010**, *49*, 2219–2222.
33. Umemoto, T.; Fukami, S.; Tomizawa, G.; Harasawa, K.; Kawada, K.; Tomita, K. *J. Am. Chem. Soc.* **1990**, *112*, 8563.
34. Ma, J. A.; Cahard, D. *J. Fluorine Chem.* **2004**, *125*, 1357.
35. Hull, K. L.; Anani, W. Q.; Sanford, M. S. *J. Am. Chem. Soc.* **2006**, *128*, 7134.
36. Dick, A. R.; Hull, K. L.; Sanford, M. S. *J. Am. Chem. Soc.* **2004**, *126*, 2300.
37. Desai, L. V.; Hull, K. L.; Sanford, M. S. *J. Am. Chem. Soc.* **2004**, *126*, 9542.
38. Kalyani, D.; Deprez, N. R.; Desai, L. V.; Sanford, M. S. *J. Am. Chem. Soc.* **2005**, 127, 7330.
39. Lewis, J. C.; Wu, J. Y.; Bergman, R. G.; Ellman, J. A. *Angew. Chem. Int. Ed.* **2005**, *45*, 1589.
40. Sun, H.; Zhang, Y.; Chen, P.; Wu, Y. D.; Zhang, X.; Huanga, Y. *Adv. Synth. Catal.* **2016**, *358*, 1946.
41. Hamashima, Y.; Yagi, K.; Takano, H.; Tamas, L.; Sodeoka, M. *J. Am. Chem. Soc.* **2002**, *124*, 14530.
42. Baudequin, C.; Plaquevent, J. C.; Audouard, C.; Cahard, D. *Green Chem.* **2002**, *4*, 584.
43. Shibata, N.; Ishimaru, T.; Suzuki, E.; Kirk, K. L. *J. Org. Chem.* **2003**, *68*, 2494.
44. Ma, J. A.; Cahard, D. *Tetrahedron: Asymmetry* **2004**, *15*, 1007.
45. Cazorla, E.; Metay, B.; Andrioletti, M. *Tetrahedron Lett.* **2009**, *50*, 3936.
46. Slocum, D. W.; Shelton, P.; Moran, K. M. *Synthesis* **2005**, 3477.
47. Anbarasan, P.; Neumann, H.; Beller, M. *Chem. Asian J.* **2010**, *5*, 1775.
48. Anbarasan, P.; Neumann, H.; Beller, M. *Angew. Chem. Int. Ed.* **2010**, *49*, 2219.
49. Yamada, S.; Gavryushin, A.; Knochel, P. *Angew. Chem. Int. Ed.* **2010**, *49*, 2215.
50. Ma, J. A.; Cahard, D. *Tetrahedron: Asymmetry* **2004**, *15*, 1007.

51. Shibata, N.; Kohno, J.; Takai, K.; Ishimaru, T.; Nakamura, S.; Toru, T.; Kanemasa, S. *Angew. Chem. Int. Ed.* **2005**, *44*, 4204.

52. Althaus, M.; Togni, A.; Mezzetti, A. *J. Fluorine Chem.* **2009**, *130*, 702.

53. Suzuki, S.; Furuno, H.; Yokoyama, Y.; Inanaga, J. *Tetrahedron: Asymmetry* **2006**, *17*, 504.

54. Suzuki, S.; Kitamura, Y.; Lectard, S.; Hamashima, Y.; Sodeoka, M. *Angew. Chem. Int. Ed.* **2012**, *51*, 4581.

55. Reddy, D. S.; Shibata, N.; Nagai, J.; Nakamura, S.; Toru, T.; Kanemasa, S. *Angew. Chem. Int. Ed.* **2008**, *47*, 164.

56. Suzuki, S.; Kitamura, Y.; Lectard, S.; Hamashima, Y.; Sodeoka, M. *Angew. Chem. Int. Ed.* **2012**, *51*, 4581.

57. Hamashima, Y.; Yagi, K.; Takano, H.; Tamas, L.; Sodeoka, M. *J. Am. Chem. Soc.* **2002**, *124*, 14530.

58. Hamashima, Y.; Suzuki, T.; Takano, H.; Shimura, Y.; Tsuchiya, Y.; Goto, T.; Sodeoka, M. *Tetrahedron* **2006**, *62*, 7168.

59. Hamashima, Y.; Suzuki, T.; Takano, H; Shimura, Y.; Sodeoka, M. *J. Am. Chem. Soc.* **2005**, *127*, 10164.

60. Suzuki, T.; Goto, T.; Hamashima, Y.; Sodeoka, M. *J. Org. Chem.* **2007**, *72*, 246.

61. Enders, D.; Huttl, M. R. M. *Synlett* **2005**, 991.

62. Beeson, T. D.; MacMillan, D. W. C. *J. Am. Chem. Soc.* **2005**, *127*, 8826.

63. Marigo, M. D.; Fielenbach, I.; Braunton, A.; Kjoersgaard, A.; Jørgensen, K. A. *Angew. Chem. Int. Ed.* **2005**, *44*, 3703.

64. Steiner, D. D.; Mase, N.; Barbas, C. F. *Angew. Chem. Int. Ed.* **2005**, *44*, 3706.

65. Kwiatkowski, P.; Beeson, T. D.; Conrad, J. C.; MacMillan, D. W. C. *J. Am. Chem. Soc.* **2011**, *133*, 1738.

66. Schulte, M. L.; Lindsley, C. W. *Org. Lett.* **2011**, *13*, 5684.

67. Andrieux, C. P.; Differding, E.; Robert, M.; Saveant, J. M. *J. Am. Chem. Soc.* **1993**, *115*, 6592.

68. Becerril, M. R.; Sazepin, C. C.; Leung, J. C. T.; Okbinoglu, T.; Kennepohl, P.; Paquin, J. F.; Sammis, G. M. *J. Am. Chem. Soc.* **2012**, *134*, 4026.

69. Qiu, S.; Xu, T.; Zhou, J.; Guo, Y.; Liu, G. *J. Am. Chem. Soc.* **2010**, *132*, 2856.

70. Mankad, N. P.; Toste, F. D. *Chem. Sci.* **2012**, *3*, 72.

71. Wu, T.; Yin, G.; Liu, G. *J. Am. Chem. Soc.* **2009**, *131*, 16354–16355.

72. Enders, D.; Potthoff, M.; Raabe, G. *J. Angew. Chem. Int. Ed. Engl.* **1997**, *36*, 2362.

73. Enders, D.; Faure, S.; Potthoff, M.; Runsink, J. *Synthesis* **2001**, 2307.

74. Less, S. L.; Handa, S.; Millburn, K.; Leadlay, P. F.; Dutton, C. J.; Staunton, J. *Tetrahedron Lett.* **1996**, *37*, 3515.

75. Nie, J.; Zhu, H. W.; Cui, H. F.; Hua, M. Q.; Jun, M. *Org. Lett.* **2007**, *9*, 3053.

76. Yamada, S.; Gavryushin, A.; Paul, K. *Angew. Chem. Int. Ed.* **2010**, *49*, 2215.

77. Faye, B.; Anna, K. K.; Forsback, S.; Anna, K.; Thomas, K. I. M. N.; Glaser, M.; Luthra, S. K.; Solin, O.; Gouverneur, V. *Angew. Chem. Int. Ed.* **2015**, *54*, 13366.

78. Davis, F. A.; Qi, H. Y. *Tetrahedron Lett.* **1996**, *37*, 4345.

79. Davis, F. A.; Qi, H. Y.; Sundarababu, G. *Tetrahedron* **2000**, *56*, 5303.

80. Weijun, F.; Guanglong, Z.; Mei, Z.; Dongfeng, H.; Baoming, J *J. Fluorine Chem.* **2009**, *130*, 996.

81. Ojima, I. *Fluorine in Medicinal Chemistry and Chemical Biology*; Wiley-Blackwell: Chichester, 2009.

82. Khan, M. O.; Lee, H.; *J. Chem. Rev.* **2008**, *108*, 5131.
83. Combettes, L. E.; Schuler, M.; Patel, R.; Bonillo, B.; Odell, B.; Thompson, A. L.; Claridge, T. D. W.; Gouverneur, V. *Chem. – Eur. J.* **2012**, *18*, 13126.
84. Laurenson, J. A. B.; Meiries, S.; Percy, J. M.; Roig, R. *Tetrahedron Lett.* **2009**, *50*, 3571.
85. Wu, J. *Tetrahedron Lett.* **2014**, *55*, 4289.
86. Hollingworth, C.; Gouverneur, V. *Chem. Commun.* **2012**, *48*, 2929.
87. Xu, W.; Wu, Z.; Wang, J. *Org. Lett.* **2016**, *18*, 576.
88. Dong, X.; Yang, W.; Hu, W.; Sun, J. *Angew. Chem. Int. Ed.* **2015**, *54*, 660.
89. Vora, H. U.; Heeler, P.; Rovis, T. *Adv. Synth. Catal.* **2012**, *354*, 1617.
90. Tao, X.; Guosheng, L. *Synlett*, **2012**, 955.
91. Lozano, O.; Blessley, G.; Martinez, T.; Amber, L.; Thompson, G. T.; Bettati, M.; Walker, M.; Borman, R.; Gouverneur, V. R. *Angew. Chem. Int. Ed.* **2011**, *50*, 8105.
92. Jiang, H.; Falcicchio, A.; Jensen, K. L.; Paixão, M. W.; Bertelsen, S.; Jørgensen, K. A. *J. Am. Chem. Soc.* **2009**, *131*, 7153.

CHAPTER 3

Microwave-Mediated Nontemplate Synthesis of Some Novel Macrocyclic Ligands for Potential Molecular Metallacages

KSHIPRA SONI, NAVEEN SHARMA, R. V. SINGH, and NIGHAT FAHMI*

Department of Chemistry, University of Rajasthan, Jaipur 302004, India

Corresponding author. E-mail: nighat.fahmi@gmail.com; rvsjpr@hotmail.com

ABSTRACT

This study addresses the development of green microwave methodologies for the preparation of a few novel nitrogen donor macrocyclic ligands. Microwave-assisted organic synthesis is a highly advantageous and eco-compatible technology for drug development. Owing to its vitality in organic and inorganic synthetic procedures, this technique has in recent times, caught the attention of researchers. Herein, we have synthesized the desired compounds using thermal and microwave-assisted nontemplate methods by the reaction of benzildihydrazone or diacetyldihydrazone and diacids (pyridine-2,3-dicarboxylic acid, pyridine-2,4-dicarboxylic acid, and pyridine-3,4-dicarboxylic acid) in 1:1 molar ratio. The use of alternative energy source as microwave makes this protocol environmentally benign. The newly synthesized ligands were characterized via spectral studies such as IR, ^{1}H NMR, ^{13}C NMR, mass, and also by elemental analysis and molecular-weight determinations. A comparative study of the ligands synthesized through microwave-assisted and conventional methods has been described.

3.1 INTRODUCTION

Recently, microwave-assisted chemical synthesis that is also recognized as a "green synthesis" emerged as a new area in the field of research in chemistry. It has been applied as a very effective way to quicken the course of many inorganic reactions because microwaves selectively absorb by polar molecules present in the reaction vessel leading to the fast rise in temperature, and as a result, the rate of reaction enhanced produces higher yield with excellent purity, lowering the quantities of side products in short period of time with easy and simple workup. The increased rate of chemical reactions on microwave exposure is due to the interaction between the sample and the electromagnetic field which leads to thermal and non-thermal effects. Nowadays, microwave has become a significant nonconventional energy source.[1–7] Macrocyclic Schiff base nitrogen donor ligands are being studied extensively owing to their mixed hard–soft donor character, diverse coordination behavior, and numerous biological activities such as toxicity against bacterial growth, anticancer, and antitumor.[8,9] They also form stable complexes with most transition metals in different oxidation states.[10] The binding selectivity of multidentate Schiff base ligands can be altered subtly by making the ligand orient donor atoms in a specific spatial arrangement.[11–14] Currently, a great deal of attention is being garnered upon macrocyclic ligands as they play a key role in various aspects of chemical and medicine industries.[15–17]

Moreover, compounds with unique physical and chemical properties are formed by the combination of the hydrazones group with other functional groups such as diacids. The benzildihydrazone macrocyclic complexes are chemically highly reactive as the chelate rings[18] contain hydrazone linkages having nucleophilic nitrogen. This enhanced reactivity of benzildihydrazone greatly widens its scope as a precursor for the preparation of a variety of bulky aromatic molecules. Schiff base condensation of dihydrazone with suitable acids would result in the formation of macrocyclic and macroacyclic molecules with a greater number of nitrogen atoms due to which they behave as better coordinating ligands for a number of metal atoms/ions. The importance of this cyclization procedure are as follows: (1) the process is a simple two-step reaction involving cheap starting materials; (2) high yields within a short time interval; (3) due to high flexibility and versatility, these polydentate ligands can form a large number of complexes with transition metals that can further be used to study the diverse coordination behavior of these ligands and also their biological applications. In view of the medicinal importance of macrocyclic compounds and in continuation of our long-term interest toward the development of newer strategy for synthesizing macrocyclic complexes[19-23] of wide variety, we herein report the microwave-assisted nontemplate eco-friendly synthesis of a new class of tetraazamacrocyclic ligands from benzildihydrazone as well as diacetyldihydrazone and different diacids using methanol as solvent and their characterization.

3.2   EXPERIMENTAL SECTION

3.2.1   GENERAL INFORMATION

Analytical grade solvents and commercially available reagents were used with no purification. Toshniwal apparatus was used for melting point determination and all melting points were uncorrected. The purity of the compounds was confirmed by TLC on silica Gel G. Molecular weights were determined by Rast's camphor method.[24] Nitrogen was estimated by Kjeldahl's method.[25] Carbon and hydrogen analyses were performed at the CDRI, Lucknow. IR spectra were recorded with the help of a Nicolet Magna FTIR-550 spectrophotometer using KBr pellets. ^{1}H NMR and ^{13}C NMR spectra were recorded in DMSO-d6 using tetramethylsilane (TMS) as an internal standard on a Bruker spectrophotometer at 400 and 100 MHz, respectively, at Saurashtra University, Gujarat. Chemical shifts are expressed in parts per million (ppm) using TMS as an internal standard. The following abbreviations were used to explain the multiplicities: (1) s = singlet, (2) d = doublet, (3) t = triplet, and (4) m = multiplet. Mass spectrum was conducted from USIC, University of Rajasthan, Jaipur. Before recording the analytical data (that agree well with the theoretical values), macrocycle compounds were first heated in a vacuum drying pistol using steam and then stored under vacuum for several days.

3.2.2   PREPARATION OF BENZYL DIHYDRAZONE/ DIACETYLDIHYDRAZONE

Dihydrazone was prepared using reported method.[22,23] A weighed amount of benzil (0.840 g, 4 mmol)/diacetyl (0.344 g, 4 mmol) was dissolved in

20 mL of methanol, and hydrazine hydrate (0.4 g, 8 mmol) was also added. The yellow solution formed was left in the air for 20 h after refluxing for 3 h. The precipitated white microcrystalline compound was collected by filtration and washed successively with 100 mL of water and 25 mL of diethyl ether. It was then dried in air. The compound was recrystallized using warm methanol and was found to be soluble in acetonitrile, diethyl ether, acetone, and chloroform (Scheme 3.1).

3.3   PREPARATION OF MACROCYCLIC LIGANDS

Synthesis of macrocyclic ligands from benzildihydrazone/diacetyldihydrazone with diacids: Two different routes were used for the synthesis of ligands.

Microwave method: A mixture of benzildihydrazone **3a** (2 mmol, 0.446 g) or diacetyldihydrazone (2 mmol, 0.22 g) **3b** and different diacids (2 mmol) {pyridine-2,3-dicarboxylic acid (0.324 g) **4**, pyridine-3,4-dicarboxylic acid (0.324 g) **6,** and pyridine-2,4-dicarboxylic acid (0.324 g) **8**} and methanol (5 mL) as solvent was taken in microwave cavity. This mixture was refluxed in microwave for 10 min. Progress of the reaction was checked after a regular interval of 1 min till the completion of reaction by TLC using chloroform: methanol (97:3) as mobile. The reaction mixture was allowed to cool at room temperature. Yellow-colored precipitates of **5a, 7a, 9a** or **5b, 7b,** and **9b,** separated out on cooling. These were filtered, washed with cold methanol, and dried under vacuum. The synthetic route for the macrocycles has been depicted in Scheme 3.2.

Thermal method: For comparison purposes, the above ligands were also synthesized via thermal method, where instead of a few drops of alcohol, the hot methanolic solution (25 mL), of benzil dihydrazone **3a** (2 mmol, 0.446 g) or diacetyldihydrazone (2 mmol, 0.22 g) **3b** and a hot methanolic solution (25 mL), of different diacids (2 mmol) {pyridine-2,3-dicarboxylic acid (0.324 g) **4**, pyridine-3,4-dicarboxylic acid (0.324 g) **6**, and pyridine-2,4-dicarboxylic acid (0.324 g) **8**} were mixed slowly in 1:1 molar ratio with constant stirring. This mixture was refluxed for 5–7 h in the presence of few drops of concentrated hydrochloric acid. The reaction mixture was allowed to cool at room temperature, and then the cooled solution was poured into ice cold water. On cooling, yellow-colored precipitates of **5a, 7a, 9a** or **5b, 7b,** and **9b**, respectively, separated out, filtered, washed with cold MeOH, and dried under vacuum. A comparison between the thermal method and microwave method is given in Table 3.1.

TABLE 3.1   Comparison Between the Microwave and Thermal Methods.

Compounds	Yield (%)		Solvent (mL)		Time	
	Thermal	Microwave	Thermal	Microwave	Thermal (h)	Microwave (m)
5a	65	80	50	5	7	6
7a	62	85	50	5	6	7
9a	70	83	50	5	6	8
5b	61	87	50	5	5	8
7b	60	90	50	4.5	6	9
9b	67	82	50	5	5	6

h, time in hours; m, time in minutes.

All the synthesized compounds were well characterized by IR, ^{1}H NMR, ^{13}C NMR, and mass analysis (**5a**). The product yields and analytical and the spectral data are given next.

(**5a**) Yellow color. m.p. 150°C. IR, v, cm^{-1} 3123 (NH), 1622 (C=N), 1695 (C=O), 1070 (N–N), 1650–1703, 1530–1580, 1255–1273, and 630–683 (CONH). ^{1}H NMR: δ 7.2312–7.9373 (m, 20H, Ar–H), 8.8425 ppm (s, 1H, NH), 8.9935 (s, 1H, NH) ppm. ^{13}C NMR (100 MHz, DMSO-d$_6$): δ ppm: 125.54, 126.08, 127.50, 127.70, 128.92, 129.00, 129.15, 129.62, 129.67, 130.80, 139.42, 141.24 (Phenyl-C), 147.17, 147.25, 147.82, 149.76, 152.69 (Pyridine-C), 166.50 (C=N), 167.67 (C=N), 191.51 (CONH), 195.21 (CONH). CHN calculated, %: C$_{42}$H$_{30}$N$_{10}$O$_4$ (738.75): C, 71.79; H, 4.82; N, 18.96. Found (738.70): C, 71.70; H, 4.78; N, 18.95. HRMS (ESI) m/z calcd for C$_{42}$H$_{30}$N$_{10}$O$_4$ (738.75) found: 738.69.

(**7a**) Yellow color. m.p. 153°C. IR, v, cm^{-1} 3125 (N–H), 1622 (C=N), 1690 (C=O), 1065 (N–N), 1650–1703, 1530–1580, 1255–1273, and 630–683 (CONH). ^{1}H NMR: δ 7.3097–7.9258 (m, 20H, Ar–H), 8.782 ppm (s, 1H, NH). ^{13}C NMR (100 MHz, DMSO-d$_6$): δ ppm 125.50, 126.52, 127.51, 128.36, 128.83, 129.25, 129.57, 131.35, 136.49, 137.77, 139.42, 141.33 (Phenyl-C), 147.15, 147.45, 147.98, 151.48, 152.40 (Pyridine-C), 166.45 (C=N), 167.89 (C=N), 191.55 (CONH), 195.2 (CONH) ppm. CHN calculated, %: C$_{42}$H$_{30}$N$_{10}$O$_4$ (738.75): C, 71.79; H, 4.82; N, 18.96. Found (738.74): C, 71.70; H, 4.80; N, 18.94.

(**9a**) Yellow color. m.p. 155°C. IR, v, cm^{-1} 3126 (N–H), 1622 (C=N), 1698 (C=O), 1067 (N–N), 1650–1703, 1530–1580, 1255–1273, and 630–683 (CONH). ^{1}H NMR: δ 7.2415–7.5370 (m, 20H, Ar–H), 8.9425 ppm (s, 1H, NH), 8.8925 (s, 1H, NH) ppm. ^{13}C NMR (100 MHz, DMSO-d$_6$): δ ppm: 125.51, 126.26, 127.67, 128.29, 128.90, 129.10, 129.36, 129.69, 131.88, 139.40, 139.59, 141.34 (Phenyl-C), 147.65, 148.53, 148.69, 150.76, 152.60 (Pyridine-C), 166.48 (C=N), 167.77 (C=N), 191.45 (CONH), 195.23 (CONH) ppm. CHN calculated, %: C$_{42}$H$_{30}$N$_{10}$O$_4$ (738.75): C, 71.79; H, 4.82; N, 18.96. Found (738.72): C, 71.72; H, 4.79; N, 18.93.

(**5b**) Yellow color. m.p. 165°C. IR, v, cm^{-1} 3111 (NH), 1639 (C=N), 1690 (C=O), 1070 (N–N), 1650–1703, 1530–1580, 1255–1273, and 630–683 (CONH). ^{1}H NMR: δ 2.26 (s, 6H, CH$_3$–C), 2.24 (s, 6H, CH$_3$–C), 8.8425 ppm (s, 1H, NH), 8.9935 (s, 1H, NH) ppm. ^{13}C NMR (100 MHz, DMSO-d$_6$): δ ppm: 31.49, 36.56 (CH$_3$), 126.24, 127.75, 139.20, 139.54 (Pyridine-C), 166.50 (C=N), 167.67 (C=N), 191.51 (CONH), 195.21 (CONH) ppm. CHN calculated, %: C$_{22}$H$_{22}$N$_{10}$O$_4$ (490.47): C, 53.87; H, 4.52; N, 28.56. Found: C, 53.84; H, 4.50; N, 28.55.

(**7b**) Yellow color. m.p. 170°C. IR, v, cm^{-1} 3112 (N–H), 1639 (C=N), 1695 (C=O), 1065 (N–N), 1650–1703, 1530–1580, 1255–1273, and 630–683 (CONH). ^{1}H NMR: δ 2.26 (s, 6H, CH$_3$–C), 2.24 (s, 6H, CH$_3$–C), 8.8425 ppm (s, 1H, NH), 8.9935 (s, 1H, NH) ppm. ^{13}C NMR (100 MHz, DMSO-d$_6$): δ ppm: 31.46, 36.53 (CH$_3$), 126.26, 127.88, 138.65, 139.68 (Pyridine-C), 166.50 (C=N), 167.67 (C=N), 191.51 (CONH), 195.21 (CONH) ppm. CHN calculated, %: C$_{22}$H$_{22}$N$_{10}$O$_4$ (490.47): C, 53.87; H, 4.49; N, 28.56. Found: C, 53.85; H, 4.49; N, 28.5.

(**9b**) Yellow color, m.p. 167°C. IR (cm^{-1}) 3114 (N–H), 1639 (C=N), 1698 (C=O), 1067 (N–N), 1650–1703, 1530–1580, 1255–1273, and 630–683 (CONH). ^{1}H NMR: δ 2.26 (s, 6H, CH$_3$–C), 2.24 (s, 6H, CH3–C), 8.8425 ppm (s, 1H, NH), 8.9935 (s, 1H, NH) ppm. ^{13}C NMR (100 MHz, DMSO-d$_6$): δ ppm: 31.20, 35.33 (CH$_3$), 127.26, 127.59, 138.25, 139.00 (Pyridine-C) 166.50 (C=N), 167.67 (C=N), 191.51 (CONH), 195.21 (CONH) ppm. C$_{22}$H$_{22}$N$_{10}$O$_4$ (490.47): C, 53.87; H, 4.49; N, 28.56. Found: C, 53.85; H, 4.49; N, 28.53.

3.4   RESULTS AND DISCUSSION

The new macrocycles were prepared by the [2+2 M] condensation reaction of the benzildihydrazone **3a** or diacetyldihydrazone **3b** with diacids **4** (pyridine-2,3-dicarboxylic acid), **6** (pyridine-2,4-dicarboxylic acid), or **8** (pyridine-3,4-dicarboxylic acid) in methanol under microwave irradiation, resulting in the formation and isolation of pure macrocyclic framework **5**, **7**, or **9**, respectively, in high yields. The physical characteristics and the analytical data of the compounds are given in experimental section. The important IR frequencies of free dihydrazones **3** and ligands formed by the condensation of dihydrazones **3** and pyridine dicarboxylic acids are present in the experimental section. The IR spectra of free benzildihydrazone/diacetyldihydrazone were compared with the spectra of the ligands. The IR spectra of the benzildihydrazone/diacetyldihydrazone display two sharp bands around 3300–3380 cm^{-1} assignable to v sym and v asym vibrations of the NH$_2$ group and a band around 3480–3620 cm^{-1} due to v (O–H) of diacids. These bands remain absent in their corresponding ligands. This clearly indicates that cyclization has taken place. Appearance of four amide bands was observed, in the regions 1650–1703, 1530–1580, 1255–1273, and 630–683 cm^{-1}, assigned to amide I, amide II, amide III, and amide IV vibrations, respectively, and similar to those reported for tetraazamacrocyclic compounds confirm the condensation of OH group of diacids and NH$_2$ group of dihydrazone and formation of macrocyclic Schiff base. Absorption bands near 1622–1639 cm^{-1} are due to v (C=N) vibrations, which is further confirmed by the absence of a signal around 3.5–4.5 ppm in the NMR spectra of the ligands corresponding to the NH$_2$ protons and appearance of resonance signal due to –CONH proton in the range of 8.74–8.99 ppm. This shows that cyclization has taken place by the loss of water molecule. The tetraazamacrocycles synthesized through the reactions presented in Scheme 3.1 are 20-membered rings with three amide nitrogen atoms (**5**, **7**, **9**). Plausible mechanism for the formation of macrocyclic ligand is depicted in Scheme 3.2.

The isolation and characterization of tetraazamacrocycles by the nontemplate method is important and interesting for studying their corresponding transition metal complexes by incorporating metal ions into the macrocyclic framework. Significantly, there have been very limited reports[26,27] of successful ring closure reactions to give metal free 20-membered "N$_4$" macrocycles, unless there are at least two secondary amine hydrogen atoms are present. These can in fact operate as an alternative to a metal ion as a "thermodynamic template"[28] in stabilizing the molecule by reducing the lone-pair repulsions.[29]

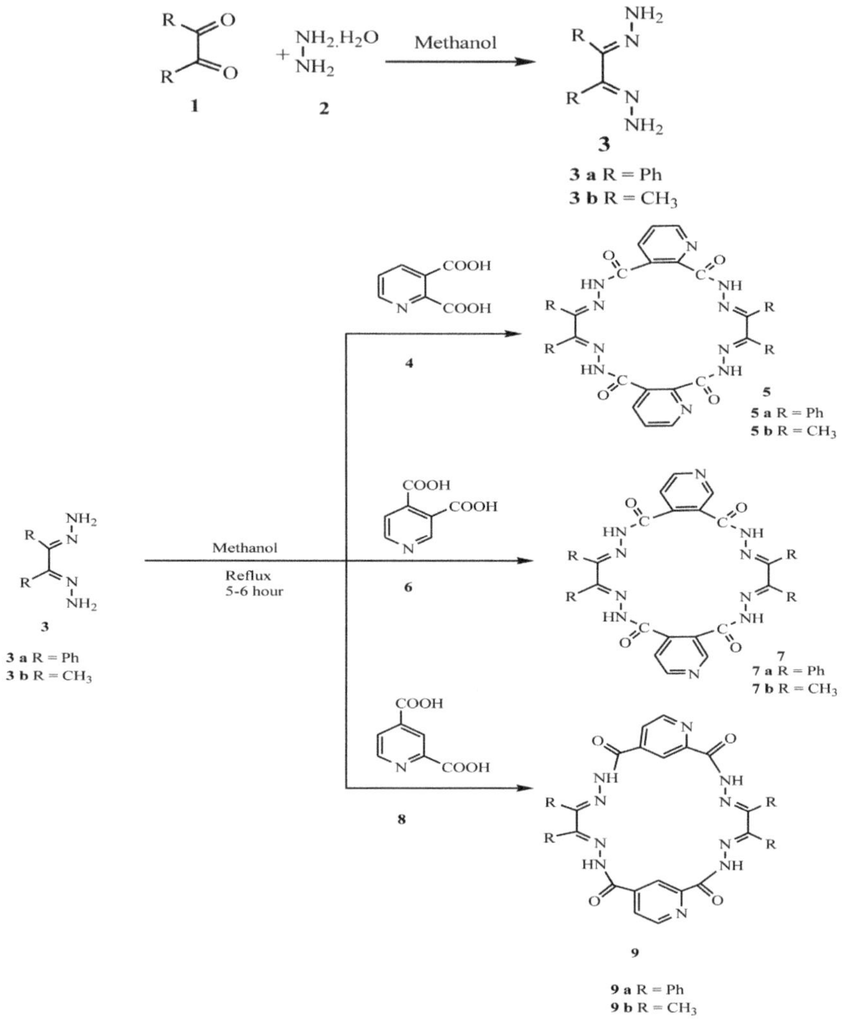

SCHEME 3.1   Synthesis of different macrocyclic ligands.

3.5 CONCLUSION

We have successfully developed a simple microwave-assisted route for the synthesis of macrocyclic cages from easily available starting materials using methanol as a solvent by green method. The characterization and two-component synthesis of tetraazamacrocyclic ligands have also been

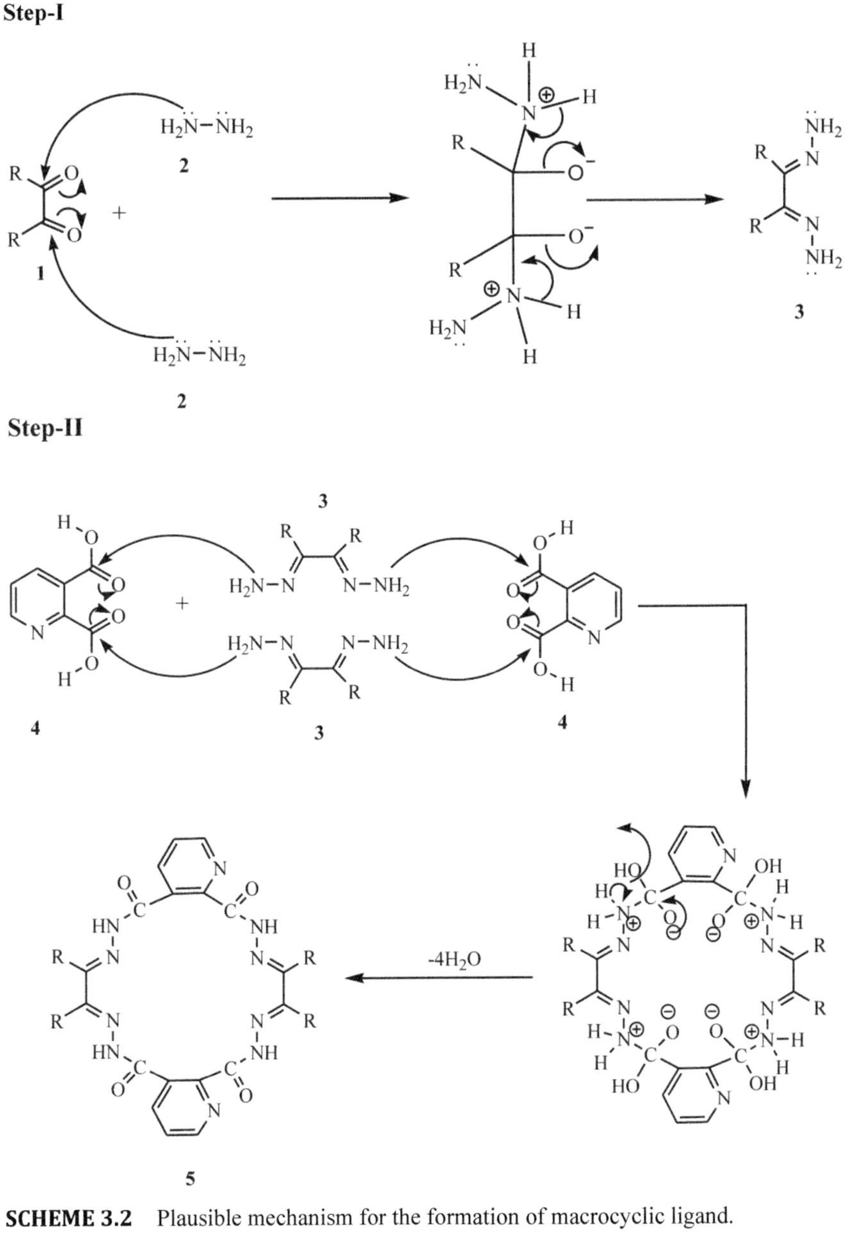

Step-I

Step-II

SCHEME 3.2 Plausible mechanism for the formation of macrocyclic ligand.

reported. This protocol is advantageous in terms of atom economy, shorter reaction time, simple, and clean reaction profiles. The newly synthesized ligands are characterized on the basis of elemental analysis, molecular-weight determinations, and spectral studies viz. IR, Mass, ^{1}H NMR, and ^{13}C NMR. The use of alternative energy source as microwave makes this protocol environmentally benign.

ACKNOWLEDGMENT

Financial assistance from the UGC, New Delhi is gratefully acknowledged through grant number F-41-4/NET/RES/JRF/1386/6190.

KEYWORDS

- **macrocyclic ligand**
- **dihydrazone**
- **diacids**
- **spectral analysis**

REFERENCES

1. Kantar, C.; Sahin, Z. S.; Buyukgungor, O.; Sasmaz, S. *J. Mol. Struct.* **2015,** *1089,* 48.
2. Srimurugan, S.; Viswanathan, B.; Varadarajan, T. K.; Varghese, B. *Org. Biomol. Chem.* **2006,** *4,* 3044.
3. Dutta, M.; Saikia, P.; Gogoi, S.; Boruah, R. C. *Steroids* **2013,** *78,* 387.
4. Surati, M. A.; Jauhari, S.; Desai, K. R. *Arch. Appl. Sci. Res.* **2012,** *4,* 645.
5. Mohanan, K.; Kumari, B. S.; Rijulal, G. *J. Rare Earths.* **2008,** *26,* 16.
6. Sun, Y.; Machala, M. L.; Castellano, F. N. *Inorg. Chim. Acta* **2010,** *363,* 283.
7. Mahajan, K.; Swami, M.; Singh, R. V. *Russ. J. Coord. Chem.* **2009,** *35,* 179.
8. Zafar, H.; Kareem, A.; Sherwani, A.; Mohammad, O.; Ansari, M. A.; Khan, H. M.; Khan, T. A. *J. Photochem. Photobiol. B Biol.* **2015,** *142,* 8.
9. Bajju, G. D.; Sharma, P.; Kapahi, A.; Bhagat, M.; Kundan, S.; Gupta, D. *J. Inorg. Chem.* **2013,** *2013,* 1.
10. Chaudhary, A.; Swaroop, R.; Singh, R. V. *Bol. Soc. Chil. Quim.* **2002,** *47,* 203.
11. Temel, H.; Ilhan, S. *Spectrochim. Acta A* **2008,** *6,* 896.
12. Lhan, S.; Temel, H.; Sunkur, M. *Indian J. Chem.* **2008,** *47,* 560.
13. Bhake, A. B.; Shastri, S. S.; Limaye, N. M. *Chem. Sci. Rev. Lett.* **2014,** *2* (6), 449.

14. Keypour, H.; Arzhangi, P.; Rahpeyma, N.; Rezaeivala, M.; Elerman, Y.; Buyukgungor, O.; Valencia, L.; Khavasi, H. R. *J. Mol. Struct.* **2010**, *977*, 6.
15. Lindoy, L. F. *The Chemistry of Macrocyclic Ligand Complexes*; Cambridge University Press: Cambridge, 1989.
16. Panda, A.; Menon, S. C.; Singh, H. B.; Morley, C. P.; Bachman, R.; Cocker, T. M.; Butcher, R. J. *Eur. J. Inorg. Chem.* **2005**, *6*, 1114.
17. Mewis, R. E.; Archibald, S. J. *Coord. Chem. Rev.* **2010**, *254*, 1686.
18. Zoubi, W. A.; Kandil, F.; Chebani, M. K. *Org. Chem. Curr. Res.* **2012**, *1*, 1.
19. Fahmi, N.; Masih, I.; Soni, K. *J. Macromol. Sci. Pure Appl. Chem.* **2015**, *52*, 548.
20. Fahmi, N.; Kumar, R.; Masih, I. *Spectrochim. Acta A* **2013**, *101*, 100.
21. Mahajan, K.; Fahmi, N.; Singh, R. V. *Indian J. Chem.* **2007**, *46A*, 1221.
22. Sharma, K.; Singh, R.; Fahmi, N.; Singh, R. V. *Spectrochim. Acta A* **2010**, *75*, 422.
23. Garg, R.; Saini, M. K.; Fahmi, N.; Singh, R. V. *Trans. Met. Chem.* **2006**, *31*, 362.
24. Melson, G. A. *Co-ordination Chemistry of Macrocyclic Compounds*; Plenum: New York, 1979; Chapter 2.
25. Owston, P. G.; Peters, R.; Ramsammy, E.; Taskar, P. A.; Trotter, J. *J. Chem. Soc. Chem. Commun.* **1960**, *980*, 1218.
26. Donaldson, P. B.; Taskar, P. A.; Alcock, N. W. *J. Chem. Soc. Dalton Trans.* **1976**, *21*, 2262.
27. Swamy, S. J.; Veerapratap, B.; Nagaraju, D.; Suresh, K.; Someshwar, P. *Tetrahedron* **2003**, *59*, 10093.
28. Vogel, A. I. *A Textbook of Organic Quantitative Analysis*; Longman: London, 2004.
29. Vogel, A. I. *A Textbook of Quantitative Chemical Analysis*; Pearson Education Ltd.; Thames Polytechnique: London, 2006; p 387.

CHAPTER 4

Lanthanide Ions–Doped Nanomaterials for Light Emission Applications

CHANDRESH KUMAR RASTOGI*, SANDEEP NIGAM, and V. SUDARSAN

Chemistry Division, Bhabha Atomic Research Centre Mumbai, Mumbai 400085, India

Corresponding author. E-mail: ckrastogi.iitk@gmail.com

ABSTRACT

Lanthanide-doped nanomaterials exhibit fascinating luminescence properties such as sharp emission, long luminescence lifetime, and emission over a wide wavelength span ranging from ultraviolet to far infrared. Exploiting these optical properties, they have been widely utilized in bioimaging, drug delivery, display devices, lasers, solid-state lighting, and solar cells. Interestingly, their luminescence characteristics are strongly correlated with the doping concentration, dopant–host combination, morphology, crystal structure, and the ligand present over the surface of the nanoparticles. Various physical methods (e.g., ball milling, physical vapor deposition, sputtering, and laser ablation) and chemical routes (e.g., chemical vapor deposition, hydrothermal, thermolysis, coprecipitation, sol–gel, and so on) are well established for the formation of lanthanide ion–doped nanomaterials. Among all, the solution-based synthesis methods or *wet chemical methods* are proven to be very promising in synthesizing the nanoparticles of desired structural features (e.g., particle size, shape, uniformity, phase, chemical composition, and functionality). Numerous wet chemical methods are well established for the synthesis of a variety of nanomaterials with each of them having some merits and demerits. A prior understanding about each of them is essential before adapting them for material preparation. In this direction, the current book chapter covers brief discussion about a few of important wet chemical methods that are very promising for the preparation of Ln-NPs of desired structural features. The chapter also deals with the basics of lanthanide ions luminescence discussing various factors that influence the luminescence characteristics of lanthanide ion–doped nanoparticles. Finally, a brief discussion about the application of luminescent Ln^{3+}-doped nanomaterials in various fields such as biomedical, lighting, and energy harvesting is also presented based on the recent literature review.

4.1   INTRODUCTION

Luminescence is the emission of optical radiation (which is not thermal in origin) from any substance upon excitation with some sort of energy. It is to be distinguished from incandescence which is the emission of radiation by virtue of the temperature of the substance (black body radiation). Luminescence is classified in different categories depending upon the source of excitation used, as an instance; if a light source is used for excitation, the phenomenon is termed as photoluminescence. Various types of luminescence processes and the related excitation sources are mentioned in Table 4.1. The list is not exhaustive and more detailed information can be found elsewhere.[1] An *Electroluminescence* is a phenomenon in which a material emits light in response to application of a strong electric field due to radiative recombination of electrons and holes in a material. The emission of light from material under excitation by a high energy electron beam (or "cathode ray") is referred to as *Cathodoluminescence*. *Triboluminescence* or *fractoluminescence* refers as the emission of light due to fracture of the crystals. It differs from piezoluminescence in which emission occurs due to deformation of the material not owing to fracture.

Luminescence-based techniques are the indispensable tool for various optical, photonic, and biological devices. Besides being used as phosphors for various biological and lighting applications, luminescent materials can be utilized as a tool to probe the materials for gathering the important information related to crystallographic defects, composition, and lattice strain. An inorganic luminescent material basically comprises few luminescent centers incorporated into a suitable host. The most commonly used luminescent centers are lanthanide ions (Ln^{3+} ions; $Ln = Nd^{3+}$, Eu^{3+}, Tb^{3+}, Gd^{3+}, Ho^{3+}, Er^{3+}, Tm^{3+}, and Yb^{3+}) and transition metal (TM) ions (e.g., Mn^{2+}, Ti^{2+}, Ni^{2+}, Mo^{3+}, Re^{4+}, and Os^{4+}) that are known as *activator* and produce variety of emissions with long luminescence lifetime. However, it is difficult to excite them as they poorly absorb the radiation. To circumvent this issue, the luminescent ions are introduced into a suitable host that acts as sensitizer and excites the activator Ln^{3+} ions by indirect pathways employing charge transfer or energy transfer mechanism (discussed next). The host material not only provides an efficient sensitization to Ln^{3+} ions but also plays a crucial role in keeping them separated so as to minimize their self-interaction–related luminescence quenching (also known as *concentration quenching*). In case the host has limited absorption capacity, another type of ions that have strong absorption characteristics (known as *sensitizer*) is codoped with activator

TABLE 4.1 Different Type of Luminescence-based on the Excitation Source Used and Their Possible Applications in Various Technologies.

S. No.	Luminescence type	Excitation source	Application
1	Photoluminescence	Photon	Fluorescent lamps
2	Electroluminescence	Electric field	EL devices, LEDs
3	Cathodoluminescence	Electrons	Field emission devices
4	Radioluminescence	X-rays or g-rays	X-ray imaging
5	Chemiluminescence	Chemical reaction energy	Analytical chemistry
6	Bioluminescence	Biochemical reaction energy	Analytical chemistry
7	Sonoluminescence	Ultrasound	Medical ultrasound risk assessment
8	Triboluminescence	Mechanical energy	Sensors

Ln^{3+} ions for their efficient sensitization. For instance, the Yb^{3+} ions have strong absorption of near-infrared radiation of wavelength ~980 nm and are used as the sensitizer for Er^{3+} ions doped in fluoride crystals. For designing an efficient luminescent material, it is anticipated that the host material must exhibit certain essential characteristics such as (1) low phonon energy, (2) favorable doping conditions for Ln^{3+} ions, and (3) improved thermal and photostability. Usually, the best hosts for these lanthanide ions are inorganic materials like crystals and glasses, because lanthanide ions generally show high quantum yields (QYs) in these hosts. It is noteworthy to mention that the luminescence properties of lanthanide-doped inorganic materials at nanoscale dimension are significantly different in comparison to their bulk counterpart. Generally, the lanthanide-doped nanoparticles (Ln-NPs) exhibit inferior luminescence characteristics; however, their processability due to improved dispersion in various organic media make them suitable for various applications. The luminescence characteristics of Ln-NPs are synergistically dependent upon several factors such as the doping concentration, dopant–host combination, crystallinity, morphology, crystal structure, and the ligand present over the surface of the nanoparticles. For the development of efficient luminescent materials, besides the selection of suitable host–guest combination, preparation of nanoparticles of desired size, shape, and crystal phase is also very crucial. Synthesis technique plays a crucial role in controlling the structural features of Ln-NPs. In past, various physical and chemical routes have been explored for the formation of lanthanide ion–doped nanoparticles. In solution-based synthesis methods, a number of process parameters, viz., temperature,[2] reaction time,[3] pH,[4] ligands,[5] additives,[6,7] and precursor salts[8–16] can be varied simultaneously, thereby providing an opportunity to tailor the atomic/molecular arrangement in a controlled fashion to produce a variety of nanoarchitecture of desired crystalline phase and narrow-particle size distribution. These methods permit the manipulation of matter at the molecular level and provide good chemical homogeneity and uniform distribution of dopant. The solution-based synthesis method has been well established for the incorporation of lanthanide ions into various hosts (such as oxides,[17] fluorides,[18] vanadate,[19] phosphate,[20] tungstate,[10,21,22] and molybdate[23]). Each of the methods has some merits and demerits associated with them. Therefore, prior knowledge about each of them is essential before adapting for nanoparticle preparation. In this prospect, the current book chapter describes a few of important solution-based synthesis methods such as hydrothermal, thermal decomposition, microemulsion, and sol–gel that are very promising to produce lanthanide-doped nanoparticles of desired chemical composition, homogeneity, and structural features. The chapter covers basics of the lanthanide ions luminescence and the various factors that influence their luminescence characteristics. Finally, a brief information is provided regarding the application of Ln^{3+}-doped nanomaterials in various fields such as solid-state lighting, drug-delivery, bioimaging, solar cells, and displays.

4.2 LANTHANIDE-DOPED INORGANIC MATERIAL

4.2.1 LANTHANIDE LUMINESCENCE

Lanthanides are the VI period elements following lanthanum in the periodic table having atomic number Z = 58–71, in which 4f shell is successively filled (Fig. 4.1). The partial filling of electrons in (4f) shell

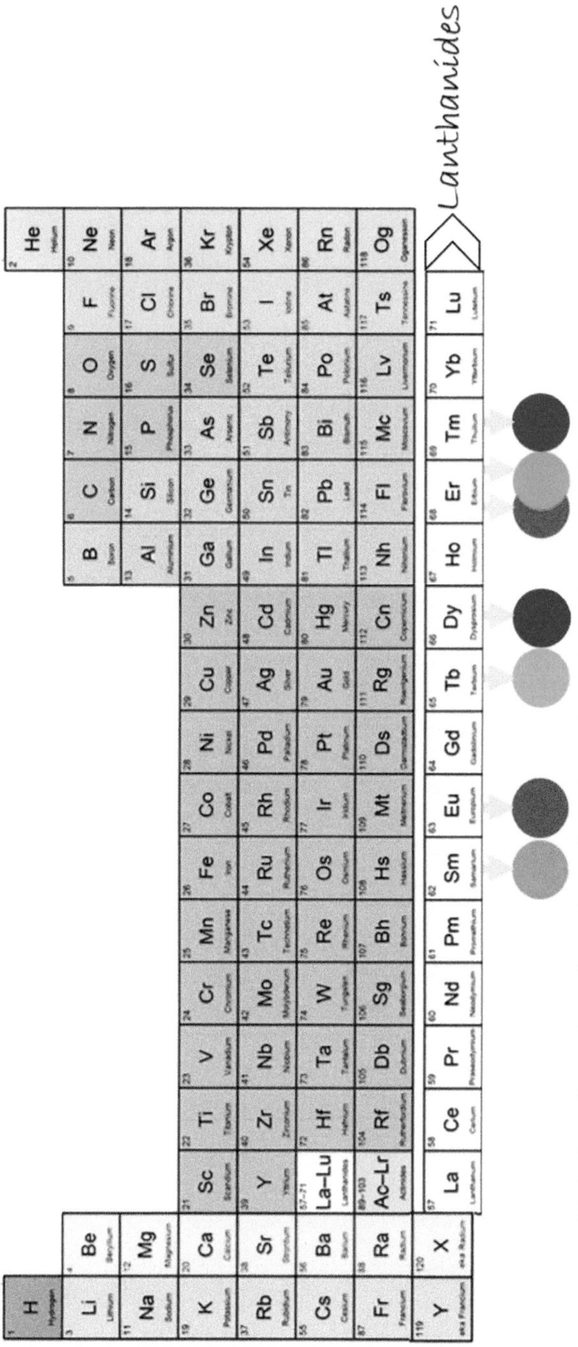

FIGURE 4.1 The periodic table of elements with the important color emission from a few of luminescent ions.

Source: Reprinted (adapted) with permission from Ref. [24]. Copyright (2017) American Chemical Society.

gives rise to useful electronic, magnetic, and optical properties. A majority of lanthanides exhibit 3+ oxidation state as the stable state with a few existing as divalent (Sm, Eu, and Yb) and tetravalent (Ce, Pr, Tb) as well.

Luminescence in lanthanide (Ln^{3+}) ions arises due to 4f ® 4f intra-configurational transitions. They exhibit sharp spectral line similar to those of free atoms or ions with high color purity. Moreover, they display low absorption and long luminescence lifetime as 4f ® 4f transitions are forbidden.[25] A typical energy level diagram of trivalent lanthanide ions is shown in Figure 4.2. The ions (Eu^{3+}, Er^{3+}, Tb^{3+}, Dy^{3+}, Tm^{3+} ions) have well-defined discrete excitation states that are responsible for sharp emissions, while a few other lanthanides, for example, Eu^{2+} and Ce^{3+}, have band-type excitation states and display broader emission spectra.

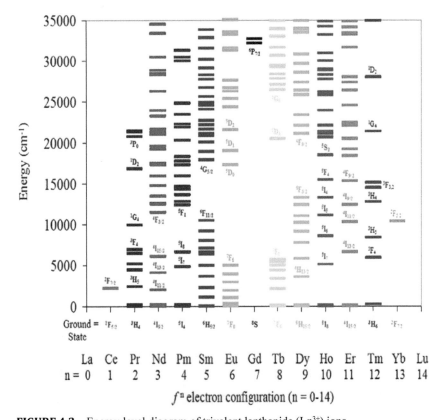

FIGURE 4.2   Energy level diagram of trivalent lanthanide (Ln^{3+}) ions.

Source: Reprinted (adapted) with permission from Ref. [40]. Copyright (2018) American Chemical Society.

Lanthanide ions–based materials display fascinating photo-physical properties such as emission ranging from ultraviolet[26] to near infrared (NIR)[26] (200–2400 nm) with long luminescence lifetime,[27–29] conventional Stokes shift emission,[30] up-conversion,[31–33] and quantum cutting[34] as depicted in Figure 4.3. Stokes shift emission is a phenomenon that involves emission of one photon of lower energy with the absorption of another of higher energy. For example, (1) Eu^{3+}, Tb^{3+}, Tm^{3+}, and Sm^{3+} ions emit red, green, blue, and orange color emission, respectively, when excited with UV light,[35,36] and (2) Nd^{3+}, Er^{3+}, Yb^{3+}, Pr^{3+}, Sm^{3+}, Dy^{3+}, Ho^{3+}, and Tm^{3+} ions produce near-infrared luminescence with irradiation of UV or visible light.[36,37] The difference of the absorption and emission energies is liberated as heat to surroundings via nonradiative pathways. Up-conversion process is associated with sequential absorption of two lower energy photons leading to emission of a photon of higher energy. For example, with an excitation wavelength of 980 nm, Er^{3+} ions in $NaYF_4$:Er^{3+}, Yb^{3+} produce green and red color emissions via two-photon up-conversion process while Tm^{3+} ions in $NaYF_4$:Tm^{3+}, Yb^{3+} produce blue color emission following two and three-photon up-conversion mechanisms.[38,39] Quantum cutting or down-conversion process entails the emission of two or more photons of lower energy by absorption of a single photon of higher energy. Consequently, the efficiency of quantum cutting is above 100%. Wegh et al. observed quantum cutting in $LiGdF_4$:Eu^{3+} due to energy transfer via cross-relation between ^6G_J state of Gd^{3+} ions and ^7F_J state Eu^{3+} ions.[34]

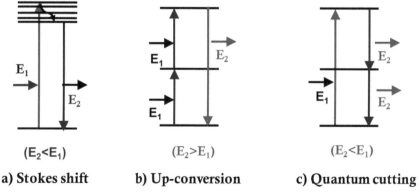

a) Stokes shift   b) Up-conversion   c) Quantum cutting

FIGURE 4.3 Schematic representation of (a) Stokes shift, (b) up-conversion, and (c) quantum cutting/down-conversion emissions.

4.2.2 SELECTION RULE AND TRANSITIONS

As stated earlier, luminescence in Ln^{3+} ions arises due to parity forbidden intraconfigurational 4f ® 4f transitions. Though these transitions are forbidden, however, they occur via mixing with $4f^N$ ® $4f^{N-1}5d^1$ transitions. Due to mixing of 4f orbitals with 5d states of opposite parity, transitions within Ln^{3+} ions are hypersensitive in nature, and a small variation in their coordination brings significant changes in the emission characteristics. This allows Ln^{3+} ions to act as a spectroscopic monitor to detect the changes of chemical environment in a complex or crystal. For example, Eu^{3+} ions can probe its local environment by measuring the relative intensity of red ($^5D_0$ ® $^7F_2$) to orange ($^5D_0$ ® $^7F_1$) color emissions, because of the hypersensitive nature of the former transition. It may be pointed out that their occurrence is strongly related to the site symmetry around Eu^{3+} ions. If the Eu^{3+} ions are located in a noncentrosymmetric environment, $^5D_0$ ® $^7F_2$ transition will dominate over $^5D_0$ ® $^7F_1$ resulting enhancement in red color emission.[41] The selection rules for different type of transitions can be summarized in Table 4.2.[42]

TABLE 4.2   Different Types of Transitions Involved in Ln^{3+} Ion Luminescence Based on *SLJ* Selection Rules.

S. No.	Type of transitions	SLJ selection rules
1	Magnetic dipole (MD) $4f^N$-$4f^N$	$\Delta S = 0$; $\Delta L = 0$; $|\Delta J| \leq 1$; $J = 0 \leftrightarrow J' = 0$ is forbidden
2	Electric dipole (ED) $4f^N$-$4f^{N-1}5d$	$\Delta S = 0$; $|\Delta L| \leq 1$; $|\Delta J| \leq 1$; $J = 0 \leftrightarrow J' = 0$ and $L = 0 \leftrightarrow L' = 0$ are forbidden
3	Judd forced (induced) ED $4f^N$-$4f^N$ (no change in parity)	$\Delta S = 0$; $|\Delta L| \leq 6$; if $L = 0$ or $L' = 0$, $|\Delta L| = 2, 4, 6$; $|\Delta J| \leq 6$; if $J = $ or $J' = 0$, $|\Delta J| = 2, 4, 6$. This implies that $J = 0 \leftrightarrow J' = 0$ and $L = 0 \leftrightarrow L' = 0$ are forbidden

4.2.3 SUITABLE HOST MATERIAL FOR LN³⁺ IONS

Doping of luminescent centers (Ln^{3+} ions) into suitable crystalline host is essentially important to obtain interesting luminescence properties. The host usually plays following key role: (1) keeps the luminescent centers away from each other to avoid the concentration quenching, (2) distributes the dopant at different crystallographic sites to produce a variety of emissions, (3) determines the local chemical environment/

symmetry, which is a key aspect for Ln^{3+} ion emission, and (4) may sensitize Ln^{3+} ions in an efficient manner. The essential criteria for the selection of suitable host materials are as follows: (1) low phonon energy, (2) favorable doping conditions, and (3) photostability. Usually, the best hosts for these lanthanide ions are inorganic materials like crystals and glasses, because lanthanide ions generally show high QYs in these hosts. For examples, the lanthanide ions–based host such as lanthanide oxides,[17] fluorides,[18] vanadates,[19] phosphates,[20] tungstates,[10,21,22] and molybdates[23] are very suitable as the desired amount of Ln^{3+} ions can be loaded into these hosts. Specifically, the fluoride-based host, especially $NaLnF_4$, is found to be most promising to produce strong luminescence due to the low phonon energy of the host.

The Ln^{3+} ions doping into nonlanthanide-based hosts is challenging if the charge and size of host cations differ significantly to Ln^{3+} ions. Few such examples are magnesium oxide (MgO), TiO_2, WO_3, ZnO, ZrO_2, and SnO_2.[43–47] The ionic radii of a few important luminescent lanthanide ions (Ln^{3+}; Ln = Ce, Nd, Sm, Eu, Tb, Dy, Er, Tm, and Yb) and some host cations such as Mg^{2+}, Zn^{2+}, Sn^{4+}, Ti^{4+}, Zr^{4+}, and W^{6+} under different anion coordination are listed in Table 4.3.

TABLE 4.3 Ionic Radius of Few Selected Cations Under Different Coordinating Environment.[54]

Element	Oxidation sate (+)	Coordination number	Crystal radius (Å)	Ionic radius (Å)
La	3	6	1.17	1.03
		7	1.24	1.10
		8	1.30	1.16
		9	1.36	1.22
		10	1.41	1.27
		12	1.50	1.36
Eu	2	6	1.31	1.17
		7	1.34	1.20
		8	1.39	1.25
		9	1.44	1.30
		10	1.49	1.35
	3	6	1.09	0.95

TABLE 4.3   *(Continued)*

Element	Oxidation sate (+)	Coordination number	Crystal radius (Å)	Ionic radius (Å)
		7	1.15	1.01
		8	1.21	1.07
		9	1.26	1.12
Mg	2	4	0.71	0.57
		5	0.80	0.66
		6	0.86	0.72
		8	1.03	0.89
	4	6	0.80	0.66
	5	6	0.76	0.62
W	6	4	0.56	0.42
	6	5	0.65	0.51
	6	6	0.74	0.60
Zn	2	8	1.04	0.90
Zr	4	8	0.98	0.84
Ti	4	8	0.88	0.74
Sr	2	8	1.40	1.26
Ca	2	8	1.26	1.12
Ba	2	8	1.56	1.42

4.2.3.1   *LOW PHONON ENERGY*

The phonon energy of the host material is associated with its lattice vibrational energy and assists in nonradiative relaxation of luminescent centers producing heat rather than emission of photons. Therefore, a material having low phonon energy is preferred for the inclusion of lanthanides to produce strong emissions. The phonon energy of various inorganic compounds follow the order fluorides < sulfides < oxides < phosphates. In this prospect, fluorides such as $NaYF_4$, LaF_3, YF_3, $LiGdF_4$ and oxyfluorides (LnOF) with a phonon energy <350 cm^{-1} are more suitable for the incorporation of luminescent Ln^{3+} ions.[48]

4.2.3.2 SUITABLE CRYSTAL STRUCTURE AND CRYSTALLOGRAPHIC SITES

Crystal structure of the Ln-NPs affects its luminescence characteristics in many ways as it determines the distribution of Ln^{3+} into the host, distance between two luminescent centers, and the coordination around Ln^{3+} ions determining local chemical environment. As an instance, for incorporation of Ln^{3+} ions in $LaVO_4$ matrix, tetragonal phase is preferred over the monoclinic counter phase as in former case the host to Ln^{3+} ions energy transfer is efficient due to the reduced separation of $[VO_4]^{3-}$ tetrahedral to Ln^{3+} ions. Further, tetragonal $LaVO_4$ offers a crystallographic site for Ln^{3+} ions that lack the inverse symmetry, thereby produces strong emissions.[49–53]

4.2.3.3 FAVORABLE CHEMICAL ENVIRONMENT

The luminescence characteristics of Ln^{3+} ions are strongly dependent on the doping conditions. Their incorporation into a suitable host may be very helpful in realizing host-mediated sensitization as well as keeping them separated to avoid concentration quenching. For their successful incorporation, it is expected that the host cations should have valency, size, and electronegativity similar to lanthanide ions. Generally, the lanthanide-based oxides, fluorides, vanadates, tungstates, and molybdates themselves met these requirements. The other nonlanthanide-based compounds containing alkaline earth ions $(Ca^{2+}, Sr^{2+}, and Ba^{2+})^{[55]}$ accept a little amount of Ln^{3+} ions as their charges are different. Similarly, doping of Ln^{3+} ions into the TM-based hosts is challenging. However, the possibility of TM to exhibit variable oxidation state may allow a small amount of Ln^{3+} incorporation despite both charge and size mismatch. The solution-based synthesis methods are more promising for the preparation of such doped system as a uniform and better mixing of dopant and host species took place at atomic/molecular levels. The substitutions of Zn^{2+},[56] Zr^{4+},[57] and Ti^{4+}[44,57] cations by Ln^{3+} ions in ZnO, ZrO_2, and TiO_2, respectively, are the few examples. However, in these hosts, the Ln^{3+} ions do not produce strong emission because of the limited amount of the luminescent Ln^{3+} ions. Therefore, it is essential that the Ln^{3+} ions must be incorporated into a suitable host with sufficient amount.

4.2.3.4   GOOD ABSORPTION CROSS SECTION

Since Ln^{3+} ions have limited absorption cross section, very often they are excited through indirect pathways, for example, sensitization by the host or codopant sensitizer ions. It is anticipated that the host must exhibit strong absorption capacity and favorable band positions relative to energy levels of Ln^{3+} ions. These features of host material provide a fair chance of host-sensitized energy transfer to excite Ln^{3+} ions. Generally, the oxides exhibit strong absorption in UV–visible region due to defect levels associated with the oxygen vacancies. While, the vanadate, tungstate, and molybdate-based host exhibits strong absorption in UV–visible spectral portion due to ligand (O^{2-}) to metal (V^{5+}, Bi^{3+}, and Mo^{6+}) charge transfer (LMCT) processes.

4.2.3.5   STABLE UNDER RADIATION EXPOSURE

The phosphor material is often used under harsh conditions, that is, exposed to different type of irradiations such as high energy photons, electrons, ions, and so forth, depending upon their application. Therefore, the host material is essentially anticipated to exhibit superior stability and withstand under irradiation. As an instance, a number of radiations are produced in the plasma display panels (PDPs) during plasma formation, which may be detrimental to the phosphor-coated over the wall of the PDP cell.

4.2.3.6   SUITABLE FOR INTENDED APPLICATION

The essential criteria for the selection of material are related to the intended application. As an instance, for the biological applications, the material must be biocompatible and stable in biological medium. In the same context, semiconductor nanocrystals are not suitable because of their toxic nature; however, they may be suitable for general lighting applications. The biocompatible host materials for the incorporation of Ln^{3+} ions are $NaLnF_4$, $GdVO_4$, $LaVO_4$, and so on.[58,59] As another illustration of other applications toward the display devices, diluted amount of Ln^{3+} ions can be introduced into MgO thin film to produce the luminescence, which may be useful for their plasma display applications. It is noteworthy to mention that MgO thin film is traditionally used as protective layer over the electrodes of the PDPs to shield the electrode from UV radiation, ions, and electrons generated during the plasma formation.[60] Fortunately, MgO absorbs vacuum ultraviolet (VUV) and UV radiations strongly due to its high bandgap ($E_g = 7.2$ eV) and inherent defects (*F*-centers), and therefore, multifunctional Ln^{3+}-doped MgO thin films can be used as protective layer as well as luminescent layer.[61]

4.3 FACTORS INFLUENCING LN³⁺ IONS LUMINESCENCE

There are various parameters that synergistically affect the luminescence properties of Ln^{3+} ion–doped into inorganic nanocrystalline material. Few of these factors are dopant–host combination, doping concentration, crystallinity, crystal structure of host, power of excitation radiation, morphology, and ligand present over the surface of nanophosphor. Each and individual factor, alone or in combination other, may influence the luminescence properties of Ln^{3+}-doped nanoparticles. The effect of each and individual factor on the photoluminescence characteristics of Ln^{3+} ions is thoroughly discussed in the available literature, and the readers are encouraged to go through the references.[14,19,31,46,48,53,62–69] In order to develop efficient luminescent materials, each of the aforementioned parameters are expected to be optimized. A few of them are discussed as follows.

4.3.1 DOPANT–HOST COMBINATION

The choice of dopant–host combination is essential in determining the luminescence characteristics of Ln^{3+} ions–based nanophosphor. The incorporation of Ln^{3+} into an inorganic host is important as it not only keeps Ln^{3+} ions separated to avoid their self-interaction but also protect it from the external environment such as solvent, ligands to avoid luminescence quenching. Further, the host may act as sensitizer and provide a necessary path way for the indirect excitation of Ln^{3+} ions *(activators)* using energy transfer or charge transfer processes. These processes are very efficient compared to their direct excitation. The energy levels of the host and dopant must be resonant enough, so that energy can be efficiently transferred to the Ln^{3+} ions. Therefore, the selection of the suitable host for Ln^{3+} ion is very crucial in determining their luminescence characteristics. A variety of luminescence such as red, green, and blue can be obtained by doping different type of Ln^{3+} ions., for example, Eu^{3+}, Tb^{3+}, and Tm^{3+} ions into a suitable host as mentioned in Table 4.4. Similarly, a single-phase pure white light-emitting phosphor can be prepared by incorporating an optimum amount of different Ln^{3+} ions, for example, a combination of Eu^{3+}, Tb^{3+}, and Tm^{3+} ions or Dy^{3+} and Tm^{3+}, and so on.

4.3.2 DOPING CONCENTRATION

Further, the luminescence characteristics of Ln^{3+} ions are greatly influenced by their doping concentration. A good density of luminescence centers is important for obtaining the bright emission; however, their number should not be increase beyond a threshold limit, otherwise may cause concentration quenching due to the decreased separation between the two neighboring centers. Therefore, it is very essential to find out an optimum doping concentration (X_c) at which the luminescence intensity is maximum. For a given Ln^{3+} ion, the value of X_c differs for host to host as it depends upon structural parameters such as cell volume and number of available sites for Ln^{3+} ions. The critical distance (R_c) between the two luminescent centers in a crystal can be expressed as,

$$R_c = 2 * (3V / 4\pi X_c N)^{1/3}$$

where X_c, N, and V are the critical doping concentration, an effective number of luminescence centers in a unit cell, and cell volume, respectively. For example, with a given value of $X_c = 0.5$, $N = 8$, and $V = 1643.9$ Å^3 in the case of Tb^{3+} ions in $K_2Y(WO_4)(PO_4)$ crystal, the critical distance between two Tb^{3+} ions is reported to be 9.21 Å, that is, the minimum separation beyond which concentration quenching may occur.

4.3.3 MORPHOLOGY

The major advantages associated with nanophosphors are related to their improved dispersion and the possibility to obtain a variety of emission by controlling their morphology. The particle size, shape, and particle-size distribution play an essential role in determining the luminescence characteristics of nanophosphor. Nanophosphors have very high surface area compare to the corresponding bulk material for a given volume. In this manner, lanthanide-doped nanoparticles (Ln-NPs) contain a large number of

TABLE 4.4 Typical Dopant–Host Combinations for Multicolored Lanthanide Ion–Doped Nanocrystals.

Dopant	Host	Crystal structure	Major emission			Refs.
			Red	Green	Blue	
Eu^{3+}	$LaVO_4$	Tetragonal	593 (W), 615 (S)	—	—	[41]
Eu^{3+}	$LaVO_4$	Monoclinic	593 (S), 615 (S)	—	—	[50]
Eu^{3+}	YPO_4	Tetragonal	593 (S), 615 (W)	—	—	[65,70]
Tb^{3+}	GdF_3	Orthorhombic	—	545 (S)	488 (S)	[71]
Tb^{3+}, Eu^{3+}	$ScPO_4$	Tetragonal	616 (S)	545 (W)	488 (W)	[72]
Tb^{3+}	$LaVO_4$	Tetragonal	—	545 (S)	488 (W)	[66]
Tm^{3+}, Tb^{3+}, Eu^{3+}	$LaVO_4$	Tetragonal	616 (S)	545 (S)	470 (S)	[66]
Ho^{3+}	Y_2O_3	Cubic	665 (M)	543 (S)	—	[73]
Ho^{3+}	LaF_3	Hexagonal	659 (S)	520 (S), 545 (S)	—	[74]
Ho^{3+}	$a\text{-}NaYbF_4$	Cubic	—	540 (S)	—	[75]
Er^{3+}	$a\text{-}NaYF_4$	Cubic	660 (S)	540 (M)	411 (W)	[76]
Er^{3+}	LaF_3	Hexagonal	659 (S)	520 (S), 545 (S)	—	[74]
Er^{3+}	$\beta\text{-}NaYF_4$	Hexagonal	656 (M)	523, 542 (S)	—	[77]
Tm^{3+}	LaF_3	Trigonal	647 (W)	—	475 (S)	[78]
Tm^{3+}	$a\text{-}NaYF_4$	Cubic	—	—	450, 475 (S)	[76]
Tm^{3+}	$\beta\text{-}NaYF_4$	Hexagonal	—	—	450, 475 (S)	[77]

luminescent centers (Ln^{3+} ions) on the surface of the nanoparticles that are greatly influenced by the surface defect, impurities, and ligands associated with Ln-NPs. The ligand effects and surface to bulk Ln^{3+} ions fraction is generally used to account for the morphology-dependent luminescence performance. Luminescence characteristics of nanophosphor can be better explained by considering the surface-area-to-volume ratio (*SA/V*) rather than just quantifying with their size and shape only. However, the prediction of *SA/V* is quite difficult. Shan et al. have explained the variation in the luminescence characteristics of $NaYF_4$:Er, Yb with the morphology (i.e., nanorod, plate, prism) in terms of S/*V* of different architecture (Fig. 4.4a). As shown in Figure 4.4b, an exponential increase in up-conversion luminescence intensity with a decrease in *SA/V* ratios is correlated with the surface properties at the interface between the nanostructure and the coordination ligands. The nanoparticles with high *SA/V* ratio (small particles) usually have more surface defects and luminescent Er^{3+} ion fraction on the particle surface, which causes the enhanced nonradiative quenching and results in the diminished luminescent intensity.

In order to obtain the improved luminescence from Ln-NPs, the surface-related nonradiative relaxation must be minimized. This can be done by the surface modification of Ln-NPs using following approaches: (1) passivation of Ln-NPs with an optically inert layer, (2) formation of core-shell nanoparticles, and (3) surface modification via ligand exchange mechanism. Chow et al. have demonstrated 30-fold increment in luminescence intensity of $NaYF_4$:Yb/Tm nanocrystals (size ~8 nm) by passivation with an inert shell of ~1.5 nm–thick $NaYF_4$. Similarly, an improvement in the QY of LaF_3:Ce, Tb from 24 to 54% is noticed by growing an optically inactive inert layer around LaF_3.[79]

The surface modification of Ln-NPs using ligand exchange mechanism may also affect the luminescence characteristics depending upon the functional group present at the surface of the nanoparticles after the ligand exchange process. Generally, the nanoparticles prepared in organic media are hydrophobic in nature due to the presence of functional group at the surface of the particle. Although the luminescence properties of hydrophilic Ln-NPs are superior, yet they are converted to hydrophilic via a ligand exchange mechanism for improving their dispersion in aqueous media. Nacchache et al. reported an improvement in the up-conversion intensity of oleic acid (OA)-capped $NaGdF_4$:Ho^{3+}/Yb^{3+} nanoparticles after ligand exchange with polyacrylic acid.

A core-shell structure is very promising in improving the luminescence efficiency of Ln-NP, as in such scheme, the core comprises Ln-NPs and shell is composed of optically inert made up of the same material without Ln^{3+} ions. This type of arrangement protects the luminescent particles and minimizes the surface-related nonradiation relaxation to improve the luminescence characteristics of Ln-NPs. Singh et al. have used this approach to introduce different red, green, and blue phosphor layer each separated by an optically inert layer using a layer-by-layer (LbL) approach for the development of white light-emitting phosphor.[66] The optically inert layer and blue, green, red phosphor layers were composed of undoped $LaVO_4$, $LaVO_4$:Tm^{3+}, $LaVO_4$:Tb^{3+}, and $LaVO_4$:Eu^{3+} nanoparticles, respectively. Such a core-shell architecture produces strong white light emission under UV irradiation (λ_{ex} = 280 nm) as shown in Figure 4.4c. This type of arrangement of phosphor layers not only minimizes the nonradiative relaxations but also avoids the cross-talking between different lanthanide ions. In such structures, the dopant Ln^{3+} ions are confined in the core of the nanoparticles, thereby suppressing the energy transfer to surface moieties and defects to result in strong luminescence. Figure 4.4d shows the digital image displaying the red, green, blue, and white light from different compositions of $LaVO_4$:Ln^{3+} (Ln = Eu/Tb/Tm/Dy) phosphors.

4.3.4 CRYSTAL STRUCTURE

As discussed earlier, the crystal structure of the Ln-NPs affects its luminescence characteristics in many ways as it determines the distance between two luminescent centers, and the coordination around Ln^{3+} ions determines local chemical environment. For example, the emission characteristics of Ln^{3+} ions in tetragonal $LaVO_4$,[53] monoclinic ZrO_2,[46] hexagonal $NaYF_4$,[81] and monoclinic $BiPO_4$[82] are improved compared to that of their respective polymorphs. These phases offer Ln^{3+} ions crystallographic sites possessing lack of inversion symmetry, which is responsible for enhanced transition probabilities leading to strong emission. As an instance, the emission intensity of Eu^{3+} ions at 615 due to hypersensitive $^5D_0 ® ^7F_2$ transition is greatly enhanced if they occupy a noncentrosymmetric site in crystal. An asymmetry parameter (A_{21}) defined as the red-to-orange luminescence intensity ratio is useful in determining the emitter site occupancy. A_{21} can be deduced by integrating the emission peaks due to an electric dipole ($^5D_0$ ® $^7F_2$) and magnetic dipole ($^5D_0$ ® $^7F_1$) transitions using the relation[83]:

$$A_{21} = \frac{\int I_2 . d\lambda}{\int I_1 . d\lambda}$$

An increase in asymmetry parameter (A_{21}) is a signature of increased occupancy of Eu^{3+} ions in noncentrosymmetric sites. Using this criterion, the Ln^{3+} ions are used as luminescent probes to find out the various important crystallographic information.[58]

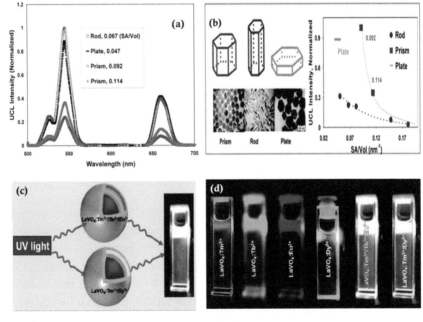

FIGURE 4.4 Morphology-dependent photoluminescence properties: (a) emission spectra of four different morphologies of up-converting $NaYF_4$:Er, Yb nanoparticles, (b) variation in up-conversion emission intensity for rod, prism, and plate type morphologies $NaYF_4$:Er, Yb nanoparticles with corresponding transmission electron micrographs,[80] (c) graphics showing the design of down-converting $LaVO_4$:Ln^{3+} (Ln = Eu, Tb, Tm, and Dy) white light-emitting phosphor, and (d) digital emission showing strong emission from the $LaVO_4$:Ln^{3+} (Ln = Eu, Tb, Tm, and Dy) nanopowders dispersed in cyclohexane.[66]

Source: (a–b): Reprinted with permission from Ref. [80]. Copyright 2010 American Chemical Society. (c–d): Reprinted with permission from Ref. [66]. Copyright 2012 American Chemical Society.

4.4 SYNTHESIS OF LANTHANIDE IONS-DOPED NANOPARTICLES (Ln-NPs)

With the advent of nanosciences and engineering, considerable attention has been given for the development of synthesis technique to produce nanostructured materials. There are two different approaches for the formation of nanoparticles, viz. top-down and bottom-up approaches. The top-down approaches use macroscopic structures (bulk material) and produce nanostructures by their disintegration into parts. These methods often utilize microfabrication methods where extremely controlled tools are utilized to cut, mill, and shape the materials into desired size and order. In contrast to top-down methods, bottom-up approaches are based on the self-assembly of atomic or molecular components to produce the nanostructures. Among all the bottom-up approaches, solution-based synthesis techniques (or wet chemical methods) are very promising in selective preparation of ultrasmall particles with diverse morphologies and suitable crystal phase of nanostructured materials. These methods have advantages in producing nanostructures with controlled defects and more homogenous composition. Further, in wet chemical reaction methods, a number of process parameters, viz., temperature,[2] reaction time,[3] pH,[4] ligands,[5] additives,[6,7] and precursor salts[8–16] can be controlled simultaneously to influence the reaction kinetics and thereby the crystal growth in a number of ways. This may provide an opportunity to synthesize a nanomaterial with diverse structural features.

Lanthanide-activated nanophosphors can be directly obtained from solution-based bottom-up approach such as sol–gel, hydrothermal, and thermal decomposition methods. In recent years, various lanthanide-doped materials such as oxides,[17] fluorides,[18] vanadates,[19] phosphates,[20] tungstates,[10,21] and molybdates[23] have been prepared by various wet chemical methods and explored for their morphology-dependent luminescence properties. A brief discussion on a few selected synthesis methods such as sol–gel, hydrothermal, coprecipitation, and thermal decomposition along with the advantages and disadvantages of each will be highlighted in this section. The scope of this chapter is limited to describe some selected solution-based synthesis methods for making nanoscale particles and is not intended to provide an exhaustive review. For further details concerning the procedures of the examples in this section and for descriptions of other related methods, the readers are encouraged to consult the references.[84,85]

4.4.1   SOLVOTHERMAL/HYDROTHERMAL METHOD

Hydro (solvo) thermal treatment is a wet chemical approach for the preparation of phosphors with controlled particle size, shape, and doping composition. This method allows the preparation of material in aqueous solutions under high vapor pressure and high-temperature condition using a reactor consisting of a steel pressure vessel called an *autoclave*. In this method, water serves two purposes as it works as a solvent for the precursors and acts as a pressure transmitting medium. The hydrothermal conditions effectively bring down the activation energy for the formation of final phase, which can also speed up the reaction; otherwise, it would occur only at a higher temperature. The unique pressure–temperature interaction of the hydrothermal solution allows the controlled preparation of different phases of various fluoride, tungstate, and a vanadate-based phosphor that is difficult to prepare with other synthetic methods.[85] The high-pressure temperature conditions improve the diffusion behavior and are very useful in the doping of alien species. The lanthanide-activated phosphors with high purity and homogeneity can be achieved under such an extraordinary environment. Figure 4.5 shows a schematic representation of a hydrothermal synthesis carried out in an autoclave to produce the nanostructures.

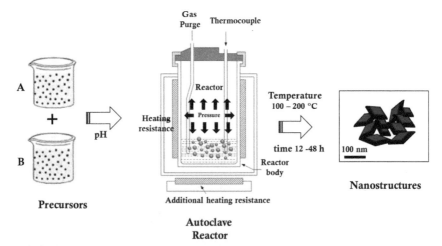

FIGURE 4.5   Schematic representation of a hydrothermal synthesis carried out in an autoclave to produce the nanostructures.

4.4.1.1 ADVANTAGES

This method offers many advantages over conventional and nonconventional synthesis methods. It has the ability to precipitate the material with good crystallinity directly from a solution that regulates the rate and uniformity of nucleation, growth, and ageing, which affects size, morphology, and aggregation control that is not possible with many synthesis processes. Unlike many advanced methods for the preparation of nanomaterials, the respective costs for instrumentation, energy, and precursors are far less for hydrothermal methods. Further, this method is more environment-friendly compared to that of many other methods. The low reaction temperatures also avoid other problems encountered with high-temperature processes such as poor stoichiometric control due to volatilization of components and stress-induced defects (e.g., microcracks) caused by phase transformations that occur as the phosphor is cooled to room temperature. The other advantages of hydrothermal synthesis are that this method can be hybridized with other processes like microwave, electrochemistry, ultrasound, optical radiation, and hot-pressing to gain benefits such as enhancement of reaction kinetics and increase ability to make novel materials. Further, this method does not need any seed, catalyst, and harmful and expensive surfactant or template; thus, it is promising for large-scale and low-cost production with high-quality crystal.

4.4.1.2 DISADVANTAGES

The method also suffers from few disadvantages such as (1) high cost of the equipment that comprises an autoclave reactor and the heating arrangement, (2) prone to explosion due to high pressure developed, therefore require careful handling with safety, (3) hydrothermal slurries are potentially corrosive, and (4) an operating temperature of more than 300°C cannot be attained.

4.4.1.3 BASIC COMPONENT AND SALIENT FEATURES OF AN AUTOCLAVE

An ideal hydrothermal autoclave must exhibit certain salient features such as it (1) must be inert to acids, bases, and oxidizing agents, (2) should be easily assembled and dismantled, (3) have sufficient length to obtain the desired temperature gradient, (4) should be leak-proof at desired temperature and pressure, and (5) able to withstand under elevated pressure and temperature conditions for a longer operation time. A hydrothermal reactor comprises an outer high-quality stainless steel jacket and inner Teflon liner or chamber. There are two different varieties of autoclave, viz, polytetrafluoroethylene (PTFE) or Teflon-lined autoclave and polypropylene (PPL) liner-based reactor. In a Teflon-lined autoclave, the maximum attainable temperature is 240C, while the safe limit is 200°C. On the other hand, the PPL-based lined autoclave can perform up to a maximum of 280°C without any deterioration. Generally, a pressure of 10–150 kbar is attained depending upon the chosen temperature and solvent.

4.4.2 COPRECIPITATION METHOD

The coprecipitation method is considered as one of the most convenient approaches for the synthesis of lanthanide-activated phosphors. The method involves the simultaneous precipitation of several metal cations in common reaction medium to form hydroxide, oxalate, and carbonate, followed by subsequent heat treatment at required temperature to produce a crystalline powder. The solubility product of each component determines its precipitation rate. In a typical synthesis procedure, solutions of metal precursors are first prepared either using suitable metal salts or by dissolution of metal oxides in an acidic solution. Thereafter, these solutions are mixed in predetermined concentration ratios to induce local super-saturation, primary nucleation, and subsequent stages of crystallization. Above a critical size, the phosphor precursors with narrow size distribution can precipitate out of the solution. The precipitation occurs under certain experimental conditions such as optimized pH, temperature, and time. The collected precipitate is then washed several times to remove the byproduct and unreacted species. The washing and drying procedures sometimes result in agglomeration; therefore, these processes must be executed carefully for the synthesis of nanoparticles with narrow-size distribution. The chemical precipitation does not proceed in a controlled way; therefore, the solids obtained by precipitation method have a wide particle size distribution and uncontrolled particle morphology with agglomerations. For obtaining the coprecipitate of well-defined stoichiometry and composition, it is essential that the product should be insoluble in the reaction medium and the precipitation kinetics must be fast enough to precipitate the product rather than getting dissolved into the reaction medium. Particular attention must be paid as the precipitation rate of different ions varies. To circumvent this issue, surfactants are introduced into the reaction medium to synchronize the coprecipitation process. The Ln-NPs are formed with organic ligands and prevent agglomeration through their adsorption to the surface of the nanoparticles. As an instance, for the precipitation of Ln-doped $LnVO_4$ particles, generally the nitrate or chloride precursors are usually adopted as a source for Ln^{3+} ions, while vanadates such as Na_3VO_4 or $(NH_4)_3VO_4$ are used to provide $[VO_4]^{3-}$ group. Yi et al. have synthesized metastable α-phase of $NaYF_4:Yb^{3+}$, Tm^{3+} up-converting nanoparticles (UCNPs) with a particle size of ~37–166 nm using precipitation method. Later, they carried out a calcination process to convert the cubic (α)-phase into to hexagonal (β)-phase, which is considered as a better host for luminescent Ln^{3+} ions. However, heat treatment process induces aggregation of nanoparticles, which limits their dispersibility in organic solvents/aqueous medium and makes them inefficient for biological applications.

4.4.2.1   ADVANTAGES

Coprecipitation process involves the atomic-scale mixing, and hence, the calcination temperature required for the formation of final product is low, which leads to lower particle size. Compared to other techniques, there is no need for costly equipment, stringent reaction conditions, and complex procedures, resulting in less time consumption. In some rare instances, crystalline nanoparticles were formed directly by coprecipitation, eliminating the need for a calcination step or postannealing process.

4.4.2.2   DISADVANTAGE

This method is not suitable if the reactants have very different solubility and precipitation rate. Further, the method is not useful for the preparation of high purity stoichiometric phase.

4.4.3   SOL–GEL METHOD

The sol–gel method involves transformation processes in which solid particles suspended in a liquid (known as *sol*) form a three-dimensional network extended throughout the liquid through hydrolysis, polymerization, gelation, and poly-condensation reactions of molecular precursors.[86] A typical sol–gel process starts with mixing of precursor salts in some solvent, usually water or alcohol at ambient or slightly elevated temperature to form a transparent solution (known as *sol*). The sol gradually evolves toward the formation diphasic system comprising a polymeric network made up of liquid and solid phase, known as a *gel*. The excess liquid phase of as obtained *gel* is then evaporated in such a way their solid network is retained. If the liquid phase is replaced by the air without disrupting its network structure, the product obtained is referred to as *aero-gel*; however, if the liquid phase is removed by evaporation in such a way that it retains the solid network structure with shrinkage of more than 90%, it results in xerogel. The method of drying will dictate whether an aero-gel or xerogel is formed. The as-obtained aero/xero gel is ground well and heat-treated at a higher temperature to produce the nanocrystalline powder. Various steps involved in the sol–gel synthesis process are described using a schematic shown in Figure 4.6.

FIGURE 4.6   Schematic diagram depicting various steps involved in the sol–gel process for the preparation of nanopowder.[87]

4.4.3.1   ADVANTAGES

The sol–gel processes serve few advantages such as, (1) they provide a better homogeneity and phase purity, (2) high-temperature sintering is not required, (3) small-sized particles (in nm range) can be prepared, (4) possibility to dope dissimilar species due to network formation, and (5) do not require any complicated equipment.

4.4.3.2 DISADVANTAGES

The sol–gel method has disadvantages in terms of poor control over the morphology and leads to aggregation of nanoparticles. Further, the several steps involved in the overall synthesis procedure make it difficult in monitoring the processes.

4.4.4 THERMOLYSIS

Thermolysis or thermal decomposition is a process of decomposition of metal cations, molecular anions, or organo-metallic compound at high temperatures. Thermolysis-based methods are very useful for the controlled preparation of nanoparticles of uniform size-shape and narrow-particle size distribution. In a typical synthesis process of fluoride-based nanostructures, metal trifluoroacetate is thermally decomposed to provide the corresponding metal and fluoride in solvents with the high boiling point such as octadecene (ODE), OA, and oleylamine (OM). A long-chained polymer such as OA is employed as ligands that control the growth of the particles by selectively adsorbing on the particular facet leading to the formation of diverse morphologies and phases. In past, thermal decomposition method is utilized for the preparation of diverse morphologies of lanthanide-doped nanocrystals.[88–93]

Zhang et al. synthesized monodispersed triangular nanoplate of LaF_3 by the thermal decomposition of $La(CF_3COO)_3$ hydrate in OA and based mixed solvents.[94] Later on, Boyer et al. have synthesized cubic phase a-$NaYF_4$ nanoparticles codoped with Yb^{3+}/Er^{3+} or Yb^{3+}/Tm^{3+} via thermal decomposition of metal trifluoroacetate precursors in the presence of OA and octadecene (ODE).[95] The noncoordinating ODE is generally used as a primary solvent due to its high boiling point while OA serves as solvent as well as a passivating ligand that prevents the nanoparticles from agglomeration. Using the same approach, later on, Boyer et al. synthesized a-$NaYF_4$:2% Er^{3+} 20% Yb^{3+} with a remarkably narrow particle size distribution by introducing the lanthanide precursors slowly into the high-temperature reaction mixture through a stainless-steel canula.[38] Table 4.5 lists various morphologies and crystal structures of different materials such as $NaYF_4$, LaF_3, Ln_2O_3, and $LuPO_4$ prepared with thermal-decomposition.

TABLE 4.5 Morphology and Crystal Phase of Various Ln-NPs Prepared Using Different Methods.

Synthesis method	Material	Solvent	T* (°C)	Shape	Size (nm)	Crystal structure	Refs.
Thermolysis	$NaYF_4$:Ln^{3+}(Ln = La to Lu each)	OA, ODE, OAM	250–330	Hexahedron, nanorod, cube, shphere, plates, prism	6–100	Cubic (α) and hexagonal (β)	[62]
	$NaYF_4$:Er^{3+}, Yb^{3+}	OM	330	Spherical		Hexagonal (β)	[96]
	$NaYF_4$:Yb^{3+}, Ln^{3+} (Ln = Er, Ho, Tm)	OA, TOPO	330–370	Nanorod, sphere, plate, cube	7–11	Hexagonal (β)	[97]
	LaF_3	OA, ODE, OM	280	Triangular, hexagonal, polygonal –nanoplate	26–55	Trigonal, orthorhombic	[94]
	$LiYF_4$	OA, ODE	290–330	Sphere, hexagon	21	Tetragonal	[26,98]
	Ln_2O_3	OA, ODE, OAM	310	Nanodisk, plate	5–16	Body-centered cubic	[99]
Hydrothermal	$LuPO_4$	–	–	Spherical nanoparticles	6–8	Tetragonal	[100]
Hydrothermal	$Ca_9Eu(PO_4)_7$	DI water	–	Nanorod	60 ′ 40	Trigonal	[101]
Sol–gel	Gd_2O_3:Eu^{3+}	DI water and ethanol (v/v = 1:1)	Calcined at 900°C	Irregular	80–140	Cubic	[102]
Precipitation	Y_2O_2S:Eu^{3+}	–	Calcined at 800°C	Nanorod	10	Cubic	[68]
Microemulsion	$SrAl_2O_4$:Eu^{2+}, Dy^{3+}	–	Calcined at 900°C	Spherical nanoparticle	40	Monoclinic	[85]

T*, synthesis temperature (°C); OA, oleic acid; ODE, 1-octadecene; OAM, oleylamine; TOPO, tri-n-octylphosphine oxide

4.4.5 MICROEMULSION METHOD

A microemulsion is a thermodynamically stable isotropic liquid mixture of nonpolar (oil), polar (water), and surfactants. The surfactant molecules form an interfacial film separating the polar and the nonpolar domains. This interfacial layer forms different microstructures ranging from droplets of oil dispersed in a continuous water phase (O/W-microemulsion) over a bicontinuous "sponge" phase to water droplets dispersed in a continuous oil phase (W/O-microemulsion). The latter can be used as nanoreactors for the synthesis of nanoparticles with a low polydispersity. The microemulsion-based method enables better control over the geometry, morphology, homogeneity, and surface area of the nanophosphor. The choice and amount of the surfactant are crucial in the formation of the emulsion and ultimately in controlling the growth process of the phosphors. Generally, the microemulsion synthesis is performed at low temperature preferably below 100°C. In a typical synthesis process of $SrAl_2O_4$:Eu^{2+}, Dy^{3+} nanophosphors by reverse microemulsion, an emulsion was first prepared by mixing an aqueous solution of nitrates with nonpolar cyclohexane and a binary surfactant mixture (polyoxyethylene-10-octyl phenyl ether and 1-hexanol). It was then slowly added into heated kerosene for water evaporation to yield phosphor precursors that are subsequently annealed at 900°C to produce high crystalline $SrAl_2O_4$:Eu^{2+}, Dy^{3+} nanophosphors. The small particle size despite the heat treatment at higher temperatures may be attributed to the confinement of the constituent cations by nanoscaled micelles that restrict the growth of the particle. This cage effect by the micelles results in the formation of nanoparticles. Similar strategy is used for the synthesis of YBO_3:Ce^{3+} phosphors by which the size of the phosphors is controlled in the range 90–189 nm by varying the volumetric ratio of water to oil.[85]

4.4.5.1 DISADVANTAGES

Due to the low synthesis temperature, the prepared nanophosphors have poor crystallinity and result in low luminescence efficiency. To circumvent this issue, a further heat treatment is required, which may lead to avoidable aggregation of the particles.

4.5   CRITICAL FACTORS DETERMINING MORPHOLOGY AND CRYSTAL PHASE OF Ln-NPs

There are a number of factors (e.g., reaction time, temperature, precursor ratio, ligands, and pH) that play a synergistic effect on the reaction kinetics to influence the nucleation and growth processes. Nanoparticles of desired particle size, distribution, shape, and crystal structure can be obtained by optimizing the aforementioned reaction parameters during the synthesis of the nanoparticles. The effect of these parameters on the structural features of Ln-NPs is briefly illustrated with a few examples in this section. The further details can be found in respective references.

4.5.1   REACTION TEMPERATURE AND TIME

The reaction temperature plays a crucial role in determining the size and shape of the nanoparticles as it regulates the rate of reaction which, in turns, determines the rate of product formation. Generally, the reactions carried out at high temperature favor the formation of bigger size particles due to an increase in the rate of reaction. In order to prepare mono-dispersed nanoparticles, aggregation of NPs must be avoided which may occur due to Ostwald ripening and oriental attachment processes. In Ostwald ripening, larger particles grow at the expense of smaller size particles while oriental attachment smaller particles merge together to form bigger size particles. Ostwald ripening is predominantly influenced by temperature and causes aggregation to minimize the surface energy. Liu et al. have produced various morphologies of $NaGdF_4$:(Ce^{3+}, Yb^{3+}, and Ln^{3+}) by varying the reaction temperature and time as mentioned in Table 4.6. The observed variation in the morphology, that is, from nanoplates to nanorods, with an increase in synthesis temperature is associated with the anisotropic growth occurring along the axial direction (c-axis) as evident from high-resolution transmission electron micrographs (HRTEM) in Figure 4.7.[103]

    The metastable phase can be arrested at a nanoscale size range due to the predominance of the surface energy of the system. The smaller size nanoparticles exhibit short-range periodicity and possess a number of dangling bonds that assist metastable phase to release its lattice strain. However, the metastable phase(s) have a tendency to move toward the stable phase. If the reaction temperature does not provide enough energy to overcome the activation barrier then the metastable phase can be retained. Therefore, under certain experimental conditions, preferably at lower synthesis temperature and using suitable ligands, metastable phase can be prepared. Liu et al. have synthesized uniform and mono-dispersed $NaGdF_4$:Er^{3+}/Yb^{3+} UCNPs and systematically investigated the role of reaction temperature and time on the morphology and phase purity.[103] The reaction time and temperature are coupled variables, if the synthesis temperature increases; under the similar experimental condition, it is possible to obtain the thermodynamically stable phase for lesser reaction time. Table 4.6 demonstrates the formation of different phase(s) and the morphologies of $NaLnF_4$ (where Ln = Gd, Ce, Yb, Tb, Eu, Sm, Dy) and $LnVO_4$ (Ln = La and Eu) NPs prepared at different synthesis temperature and reaction time. A higher reaction temperature of 280°C results in formation of the thermodynamically stable hexagonal phase of $NaLnF_4$ particles despite the synthesis is carried out for a shorter reaction time of 15 min. On the other hand, a lower synthesis temperature of 200°C and a reaction time of 5 h produce metastable cubic phase of $NaLnF_4$. Similarly, the monoclinic, tetragonal, and a mixture of both the phases of $LnVO_4$ can be selectively prepared by varying the reaction temperature and time as mentioned in Table 4.6.[51]

4.5.2 LIGAND, ADDITIVES, AND SOLVENTS

The effect of ligand, surfactants, and additives on the growth of LnNPs is extensively studied in past to produce the nanoparticles of different size, shape, phase, and narrow size–distribution.[52,90] Ligands are known to form a complex with the precursor metal ions and slow down the nucleation process. Further, they can preferentially adsorb over the high-energy facets of newly formed nuclei and retard their growth. The control over the growth of the particular facet may provide an avenue to develop various morphologies. The interaction of the ligand with the surface is controlled by solvent polarity and pH of the reaction medium. Generally, the long-chain surfactants such as an OA, cetyltrimethyl ammonium bromide (CTAB), polyvinylpyrrolidone (PVP), and ethylenediaminetetraacetic acid (EDTA) are used as ligands for the controlled synthesis of Ln-NPs in organic media. As an instance, OA preferentially binds on the particular facets of the $NaGdF_4$:Ln^{3+} (Ln = Ce and Yb) nanocrystal exclusive of 001 type facets, hindering the growth in all other directions except to that of 001 type planes, resulting in epitaxial growth along 001 direction to produce nanorods-shaped particles.[103]

TABLE 4.6 Effect of Reaction Temperature and Time on the Final of Morphology of Ln-NPs.

Material	Temperature (°C)	Time (h)	Phase	Shape	Other common experimental condition
NaLnF$_4$ (Ln = Gd, Ce, Yb, Tb, Eu, Sm, Dy)	200	5	Cubic	Irregular	*Synthesis route:* Thermolysis [103]
	230	5	Cubic	Irregular	*Precursors:* Ln(oleate)$_3$
	260	5	Cubic + hexagonal	Irregular	
	280	1/4	Hexagonal	Nanoplates	Gd:Ce:Ln
	280	1	Hexagonal	Nanorods	Ln = Tb/Eu/Sm/Dy
	280	2.5	Hexagonal	Nanorods	85:10:5
	280	5	Hexagonal	Nanorods	
	280	10	Hexagonal	Spherical	
LnVO$_4$ (Ln = La and Eu)	90	48	Monoclinic	Nanoparticle	*Synthesis route:* Hydrothermal [51] *Precursor:* La(NO$_3$)$_3$ and NaVO$_3$
	120	48	Monoclinic	Nanoparticle/rod	
	150	48	Monoclinic + tetragonal	Nanoparticle/rod	
	180	48	Tetragonal	Nanorod	
	180	48	Tetragonal	Nanorod	
	180	6	Monoclinic + tetragonal	Nanoparticle/rod	
	180	24	Tetragonal	Nanorod	
	180	48	Tetragonal	Nanorod	
	180	60	Tetragonal	Nanorod	

The formation of irregular morphologies of the NPs prepared using low concentration of OA and the variations in morphology of monodispersed particles with varying ligand content clearly demonstrates the significance of ligand in selective preparation of distinct morphologies.[103]

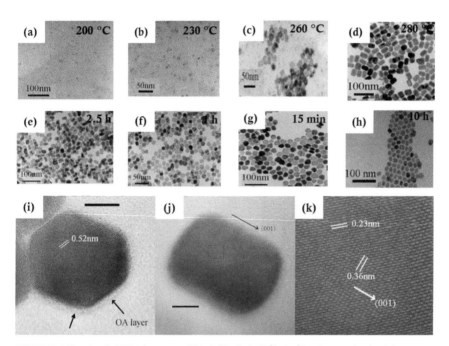

FIGURE 4.7 (a–d) TEM images of NaGdF$_4$:Ce(Yb)$^{3+}$, Ln^{3+} NCs synthesized in 15 mL oleic acid/15 mL 1-octadecene treated for 5 h at 200, 230, 260, and 280°C, respectively; (e–h) TEM images of NaGdF$_4$:Ce(Yb)$^{3+}$, Ln^{3+} NCs synthesized in 15 mL oleic acid/15 mL 1-octadecene treated at 280°C for 2.5 h, 1 h, 15 min, and 10 h, respectively. (i) HRTEM image of the top/bottom surface of a single nanorod in the image (d). The oleic acid layer (indicated by arrows) adsorbed round the side surfaces can be clearly observed; (k) HRTEM image of the side surface of the nanorod as in image (i); (k) magnified picture of the selected area in image (j). The scalar bars stand for 10 nm in images (i) and (j)[103] The details of the other experimental conditions are given elsewhere.[103]

Source: (a–h): Reprinted with permission from Ref. [103]. Copyright 2008 Royal Society of Chemistry.

Solvents influence the rate of the reaction; however, they too don't chemically react with precursor but serves as reaction medium and influences the reaction rate. The reaction will proceed in a controlled fashion if the synthesis is carried out in a nonpolar organic solvent and the precursors are introduced in a reaction medium using a polar solvent. In past, the organic solvents with high boiling point (e.g., 1-octadecene, oleylamine, and OA) and their mixtures are generally used as a solvent for carrying out the size and shape-controlled synthesis of nanomaterials. Shan et al. have produced different morphologies (e.g., spherical, rectangular parallelepiped, and cuboidal) of $NaYF_4$ with cubic (α), hexagonal (β), and a mixture of both ($\alpha + \beta$) phase(s) by varying the ratio of the solvent.[104] As mentioned in Table 4.7, α-$NaYF_4$ is formed when either of TOP or OA is used solely while β-$NaYF_4$ starts forming with an increase of the relative amount of TOP. This suggests that the $\alpha \rightarrow \beta$ phase transition was most probably caused by a ligand formed between the OA and the Lewis base TOP, which produced totally different coordination properties to affect the nucleation and growth processes.

TABLE 4.7 Effect of Solvent on the Morphology and Crystal Structure of $NaYF_4$:Ln nanocrystals.[104]

Material	Tempera-ture/time (°C/1 h)	Solvent	Morphology	Crystal structure	Other common experimental conditions
	320	OA	Spherical	Cubic	*Synthesis method:*
$NaYF_4$:33% Yb, 3% Er	320	OA/TOP 4:1	Cuboidal	Cubic + hexagonal	Thermolysis [104] Reaction
	320	OA/TOP 1:1	Cuboidal + hexagon	Hexagonal	temperature: 320°C/1 h
	320	OA/TOP 1:4	Rectangular parallelepiped	Hexagonal	
	320	TOP	Spherical	Cubic	

　　Besides morphology control, the ligand is also useful in polymorph selection, that is, selectively precipitating different phases of the material. The ligand acts as a strong chelating agent and provides a hindrance to the reacting ions to force them to crystallize in a manner with fewer coordination number (CN). As an instance, the metastable $LaVO_4$ having zircon-type tetragonal phase has been a synthesis in past using EDTA, CTAB, and OA ligands. Addition of these ligands prefers the formation of metastable zircon-type tetragonal phase of $LaVO_4$ in which La^{3+} is coordinated with eight oxygen anion (CN = 8) instead of nine in the case of thermodynamically stable monazite-type monoclinic phase (CN = 9).

4.5.3 PRECURSOR SALTS

Precursor salts can also regulate the growth of nanostructures in which the ionic character of precursor species and their steric/electronic effect determines the growth rate of different facets of the nanocrystals.[9–16] Table 4.8 lists different morphologies and crystal phase of few lanthanide ion–doped fluoride and oxides (e.g., EuF_3, $NaYbF_4$, Ce_2O_3, TiO_2, $NaLa(WO_4)_2$, and $CaMoO_4$) obtained by varying precursor salts of lanthanides. Wang et al. have utilized a solution-based synthesis method to produce a variety of EuF_3 nanostructures such as nanowire, rods, and bundles by varying precursors of fluorides, viz., caesium, rubidium, and sodium, respectively.[9] Rastogi et al. reported formation of distinct morphology of Eu^{3+}-doped sodium lanthanum tungstate phosphors, viz, nanoneedles, nanocuboids, and rugby shape nanocrystal using chloride, nitrate, and acetate precursor salts of lanthanides, respectively. The variation in the morphology was correlated to the preferential development of low index (100) and (001) facets caused by differential adsorption of Na salts of precursor anions, that is, Cl^-, NO_3^-, CO_3^{-2} which acts as growth hindering species for these crystallographic facets.[105] Fan et al. have reported the variation in phase transformation behavior of $LaVO_4$ synthesized by hydrothermal method using two different precursor salts, that is, nitrate and chloride precursor salts of La. The XRD patterns of $LaVO_4$ prepared under different synthesis conditions as shown in Figure 4.8A and B suggest that choice of precursor salt is crucial for selective crystal phase besides the reaction temperature, time, and pH of reaction medium.[51]

4.5.4 PH PARAMETER

The pH parameter is a measure of the acidic or basic condition of a solution. The variation in pH value may affect the nucleation and growth processes by controlling the surface energy of the system. An increase in pH of reaction medium decreases the surface charge density by desorption of protons, which consequently increase the surface energies of the system under consideration. Fan et al. has studied the effect of pH and precursor salt on the crystal growth and produced various morphologies of $LaVO_4$ nanoparticles with metastable zircon-type tetragonal crystal structure.[51] Table 4.9 displays the morphologies and crystal phase(s) of $LaVO_4$ produced by varying pH parameter and precursor salts under hydrothermal reaction conditions. The variation in the crystal structure of $LaVO_4$ prepared using nitrate and chloride precursor salts under different pH conditions is displayed in Figure 4.8C and D, respectively.

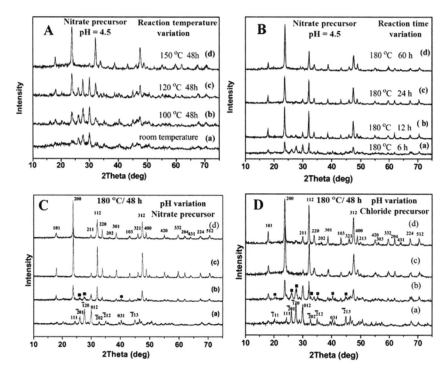

FIGURE 4.8   Effect of reaction temperature, time, pH and the lanthanide precursor on the crystal structure of LaVO$_4$ system. XRD patterns of LaVO$_4$ (A) as prepared and annealed for 48 h at (a) 100°C, (b) 120°C, and (c) 150°C; (B) annealed at 180°C for (a) 6 h, (b) 12 h, (c) 24 h, and (d) 60 h; (C) and (D) prepared at different pH (a) 2.5, (b) 3.5, (c) 4.5, and (d) 6 using nitrate and chloride precursor of lanthanides, respectively.[51]

Source: Reprinted with permission from Ref. [51]. Copyright 2006, American Chemical Society.

4.5.5   DOPANT/IMPURITY

The addition of foreign species into the host lattice may affect the growth and crystallization of nanoparticles in many ways. Very often it is found that the dopant segregates over the grain boundaries to prevent the growth of the crystals. In many cases, the dopant may induce crystallization of particular phase, which is difficult to produce under certain experimental

TABLE 4.8 Effect of Precursors, Dopant, and Ligand on the Morphology and Phase of Lanthanide Ions Containing Inorganic Materials.

Material	Precursor	Dopant	Ligand	Morphology	Crystal structure	Synthesis method
EuF_3	KF	—	—	Nanoplates	Hexagonal	Precipitation [9]
	HF			Nanospheres	Hexagonal	
	NaF			Nanobundles	Orthorhombic	
	RbF			Nanorods	Orthorhombic	
	CsF			Nanowires	Orthorhombic	
$NaYbF_4$	NaF	Er^{3+}	Trisodium citrate	Prism	Hexagonal	Precipitation [13]
	NH_4F			Prism	Hexagonal	
	$NaBF_4$			Microsphere	Cubic	
Ce_2O_3	Lanthanide nitrates, urea, glycerol	$0\% Yb^{3+}$	Oleic acid, oleylamine	Sphere	Cubic	Thermolysis [106]
		$12\% Yb^{3+}$		Cube	Cubic	
TiO_2	$C_{16}H_{36}O_4Ti$, Europium nitrate, butanediol,	$0\% Eu^{3+}$	—	Sphere $d = 58$ nm	Rutile	Sol–gel [107]
		$10\% Eu^{3+}$	—	Sphere $d = 12$ nm	Anatase	
$NaLa$ $(WO_4)_2{:}Eu^{3+}$	Chloride	$5\% Eu^{3+}$	Oleic acid	Nanoneedles	Tetragonal	Precipitation [105]
	Nitrate			Nanocuboids		
	Acetate			Rugby NCs		
	Carbonate			Rugby NCs		
$CaMoO_4$	Nitrate salts of Ca and Eu, Na_2MoO_4, Li_2MoO_4 or K_2MoO_4	Eu^{3+}, M^+ ($M =$ Li, Na, K)	No PDDA	Irregular	Tetragonal	Hydrothermal [69]
			PDDA (1 g)	Sphere $d = 1.4$ m		
			PDDA (2 g)	Sphere $d = 2.2$ m		

$Na_2H_2L\cdot2H_2O$; ethylenediaminetetraacetic acid disodium; OA, oleic acid; ODE, octadecylene; OM, oleylamine; MCs, micro crystals; Nc, nanocrystals; $C_{16}H_{36}O_4Ti$, tetrabutyl-orthotitanate; PDDA, poly-(diallyldimethylammonium chloride).

TABLE 4.9 Effect of Precursor and pH on the Final Morphology and Crystal Structure of the Nanoparticles.

Materials	Precursor	pH	Morphology	Structure	Other common experimental conditions
	$LaCl_3$	2.5	Nanoparticles	Monoclinic	Synthesis method used: hydrothermal [51]
	$LaCl_3$	3.5	Nanoparticles/rods	Monoclinic + tetragonal	Reaction temperature and time = 180°C for 48 h
	$LaCl_3$	4.5	Nanorods	Tetragonal	
$LaVO_4$	$LaCl_3$	6.0	Nanoparticles	Tetragonal	
	$La_2(SO_4)_3$	2.0	Nanorods	Tetragonal	
	$La_2(SO_4)_3$	3.0	Nanorods	Tetragonal	
	$La_2(SO_4)_3$	4.5	Nanowhiskers	Tetragonal	
	$La_2(SO_4)_3$	6.5	Nanoparticles	Tetragonal	
	$La(NO_3)_3$	4.5	Nanorod	Tetragonal	
	NH_4F	3.0	Prismatic microrods	Hexagonal	Synthesis method: hydrothermal [108]
$NaYF_4$		10.0	Hexagonal microprism	Hexagonal	Reaction temperature and time = 180°C for 24 h
	NaF	3.0	Irregular	Hexagonal	
		10	Hexagonal microprism with protruding centers	Hexagonal	

conditions. For example, the tetragonal phase of $LaVO_4$ can be arrested despite of being metastable in nature by doping smaller size Ln^{3+} ions by substituting Eu^{3+} ions at lanthanum sites. The dopant Eu^{3+} ions being smaller in size prefer a lower coordinating environment, thereby preferring to retain tetragonal phase and inhibits the formation of monoclinic phase. It is noteworthy to mention that $LnVO_4$ with bigger size cations (e.g., Ln = La, Ce, and Pr) requires coordination of nine oxygen anions for Ln^{3+} ions and thereby prefers to crystallize in monoclinic $LnVO_4$ which is the thermodynamically stable state. In contrast, for $LnVO_4$ with smaller size cations (e.g., Ln = Y, Eu, and Tb), it is preferable to have a coordination of eight oxygen anions around Ln^{3+} ions and thereby crystallizes to form tetragonal $LnVO_4$. Using these facts, the metastable phase of $LaVO_4$ has been stabilized in past by substituting smaller Eu^{3+} ions at lanthanum sites ($r_{Eu}^{3+} = 1.07$ Å and $r_{La}^{3+} = 1.22$ Å).[109]

4.6 APPLICATIONS OF LN³⁺-DOPED NANOPHOSPHORS

The use of lanthanide-doped nanoparticles (Ln-NPs) to convert photons of one wavelength to other wavelength is well-known to produce a variety of fascinating luminescence properties which offers their wide range of applications in fluorescent tubes, lasers, white light-emitting diodes (LEDs), and solar cells. Figure 4.9 demonstrates different possible applications of lanthanide ion–doped nanophosphors in various fields. Further, their fascinating luminescent properties and nontoxic behavior propose their promising application in various biomedical techniques. Herein, a few of the possible applications of typical Ln^{3+}-doped nanophosphors have been summarized in Table 4.10 and is discussed in brief as follows:

4.6.1 LIGHTING

Solid-state lighting based on white LEDs is an attractive solution for next-generation illumination, due to its outstanding energy efficiency. The development of a single-phase white light–emitting material with superior luminescence characteristics is a big challenge for current research in the field of luminescent material. Luminescent lanthanide-doped nanoparticles (Ln-NPs) can be easily dispersible in organic media and coated over the UV/blue LEDs for the fabrication of phosphor-converted LEDs.

The advantages associated with Ln-NPs for such application are that the functionalized Ln-NPs may exhibit better dispersion in organic media and can be processed easily for device fabrication. A single-phase pure white light-emitting phosphor can be prepared by incorporating an optimum amount of different Ln^{3+} ions, for example, a combination of Eu^{3+}, Tb^{3+}, and Tm^{3+} ions or Dy^{3+} and Tm^{3+}, and so on.

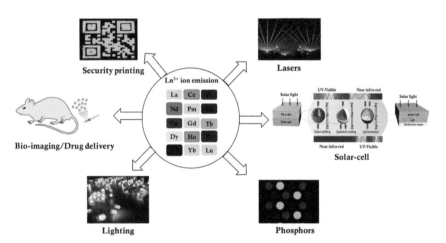

FIGURE 4.9 Schematic demonstrating various possible applications of lanthanide-doped luminescent materials.

An alternative and most popularly used approach for producing white LEDs (W-LEDs) involves a combination of a blue LED with either a yellow-green phosphor or a red-green phosphor. The phosphor materials play a key role in developing bright white light LEDs as their performance directly determines color quality, luminous efficacy, and reliability. A commercial pc-LEDs nowadays uses a blue GaN LED chip and a yellow light-emitting YAG:Ce^{3+} phosphor material. Another alternative to produce white light involves the usage of both yellow and green phosphors. In that case, CaAlSiN$_3$:Eu^{3+} can be used as a red phosphor along with YAG:Ce^{3+}-based yellow phosphor.[115] Using the phosphor blend of yellow emitting YAG:Ce^{3+} and red-emitting K$_2$TiF$_6$:Mn^{4+} phosphor, Zhu et al. have fabricated a high-performance warm-white LED.[116] The rare-earth Eu^{2+} ions are used as a popular broadband blue color emitter because its $4f^65d^1 \rightarrow 4f^7$ emission transition is sensitive to local crystal field variations and can thus be manipulated by the choice of the host material.

TABLE 4.10 Important Information Related to Few Important Hosts for Ln^{3+} Ions, Characteristic Emission of Ln^{3+} Ions and Involved Transition, and Their Possible Application.

Sample	Luminescent center	Characteristics emission (nm)	Transitions involved	Application	Refs.
Oxide	Eu^{3+}	615	$^5D_0 \circledR\ ^7F_2$	Phosphor	[110]
Fluoride	Er^{3+}, Yb^{3+}	656	$^4F_{9/2} \circledR\ ^4I_{15/2}$	Drug delivery	[111]
Ortho-vanadate	Tb^{3+}	545	$^5D_4 \circledR\ ^7F_6$	Bioimaging	[112]
Ortho-phosphate	Tm^{3+}, Yb^{3+}	470	$^1G_4 \circledR\ ^3H_6$	Photocatalyst	[113]
Molybdate	$Gd_2(MoO_4)_3$	545	$^5D_4 \circledR\ ^7F_6$	Solar cell	[114]

4.6.2   BIOIMAGING

In particular, the up-converting lanthanide-doped nanoparticles (UCNPs) exhibits excellent luminescent properties, such as large anti-Stokes shifts, low autofluorescence background, strong emission with high penetration depth. Besides, they exhibit low toxicity and better photostability, therefore becoming a suitable alternative to organic fluorophore and quantum dots (QDs) which suffers from photo-bleaching and toxicity related issues, respectively.[48,117–119] The organic dyes and QDs generally produce luminescence via down-conversion process (i.e., emit light of higher wavelength by absorbing high energy UV or visible radiation). The absorption of the light by the UV sensitive biological samples often induces autofluorescence, which causes a low-signal-to-noise ratio for fluorescence. Also, the prolonged exposure of UV radiation is detrimental to the biological samples. These issues can be circumvented by developing an up-converting nanophosphor by incorporating suitable Ln^{3+} ions (e.g., Er^{3+}, Yb^{3+}, Tm^{3+}, Ho^{3+}) into a biocompatible host such as $NaYF_4$,[118] $GdVO_4$,[59] $LaVO_4$,[58] and so forth. The following features of UCNPs make them suitable for biomedical as well as diagnostic applications such as biolabels, cell imaging and targeting, bioassays, and biosensors: (1) exhibit excellent luminescent properties and high photostability, (2) require NIR radiation for excitation which is beneficial in minimizing radiation-induced tissue damage and autofluorescence, (3) exhibit large conventional anti-Stokes shift, which allows to differentiate between excitation and emission signals, and (4) are biocompactible and nontoxic to a broad range of cell lines. Further, lanthanide-doped UCNPs find potential applications in various other multimodal bioimaging techniques (e.g., X-ray computed tomography imaging, magnetic resonance imaging, positron emission tomography imaging, single-photon emission computed tomography imaging) and several types of photo-, chemo-, and photothermal therapies.[119]

4.6.3   ENERGY HARVESTING

The main energy losses occurred during solar energy to electricity conversion are due to spectral mismatch: that is, low energy photons are not absorbed by a solar cell while high energy photons are not used efficiently. The use of lanthanide ion–doped nanomaterials to convert photons of one wavelength to different wavelengths (of interest) can be potentially exploited for the energy harvesting application.[120] The rich and unique energy level structure arising from the $4f^N$ inner shell configuration of the trivalent lanthanide ions gives a variety of options for efficient up- and down-conversion emissions (discussed earlier), which can provide a viable option to reduce the spectral mismatch. For this, both the up-converting and down-converting lanthanide-doped nanoparticles can be employed in the light-absorbing layer of solar-cell for efficient energy harvesting. In the case of up-conversion, two low-energy infrared photons that cannot be absorbed by the solar cell are added up to give one high-energy photon that can be absorbed. In the case of down-conversion, one high energy photon is split into two lower energy photons that can both be absorbed by the solar cell. Kumar et al. have studied the luminescence properties of $Gd_2(MoO_4)_3$:Eu^{3+}/Tb^{3+}/Tm^{3+} (down-shifting) and $Gd_2(MoO_4)_3$:Er^{3+}/Yb^{3+} (up-converting) nanophosphors and proposed that could be futuristic promising broad spectral converter phosphor which may possibly integrate with the next-generation Si-solar cell to enhance the efficiency of the cell. A possible combination of such a down-shifting and up-converting lanthanide ion–based nanophosphor for solar-cell application is displayed in Figure 4.10 along with the solar spectrum.[114]

4.7 SUMMARY

Lanthanide (Ln^{3+}) ions-doped nanoparticles (Ln-NPs) exhibit unique photo-physical processes such as up-conversion, down-shifting, and quantum cutting. These fascinating routes can be exploited to produce a variety of emission over the broad spectral region ranging from 200 to 2400 nm. The luminescence properties of Ln-NPs are strongly dependent on various factors such as dopant–host combination, doping concentration, crystal structure, morphology, and ligand present over the surface of the nanoparticles, and so forth. Therefore, to obtain the desired luminescence characteristics, these parameters must be optimized. The synthesis method plays a major role in determining the composition, phase, purity, crystal structure, particle size-shape, and particle size distribution of lanthanide-doped nanoparticles. Wet chemical synthesis routes, such as hydrothermal, thermolysis, microemulsion, are found to be very promising for the controlled preparation of desired size, shape, phase, and composition. In such methods, a number of parameters such as pH, reaction temperature/time, and ligands can be varied simultaneously

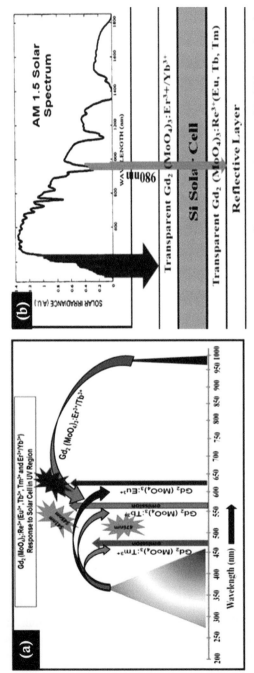

FIGURE 4.10 (a) and (b) demonstrate the schematic diagram of proposed $Gd_2(MoO_4)_3$:Ln^{3+} (Ln = Eu, Tb, Tm, and Er/Yb) nanophosphor (downshifts and up-conversion) as a broad spectral converter from UV to NIR in order to enhance the efficiency of the Si solar cell.[114]

Source: Reprinted with permission from Ref. [114]. Copyright 2015 Royal Society of Chemistry.

so as giving the possibility to control the reaction kinetics in many ways to produce a variety of nanoarchitectures and crystal phase(s). The prepared Ln-NPs exhibits excellent dispersion in various organic and aqueous media due to their nanosize effect and the functional group present at the surface of the nanoparticles, thereby found suitable for various biomedical applications. The luminescent lanthanide-doped nanoparticles can be used as phosphors for various display and solid-state lighting devices. Further, the up-converting and down-converting lanthanide-doped nanoparticles can be employed in the light-absorbing layer of solar cell for efficient energy harvesting applications. The nontoxic behavior and excellent luminescent characteristics of UCNPs propose their potential applications in various bioimaging techniques, drug delivery, and therapies.

KEYWORDS

- **lanthanide ions**
- **wet chemical methods**
- **morphology**
- **crystal structure**
- **luminescence**
- **phosphor**

REFERENCES

1. Vij, D. *Luminescence of Solids*; Springer Science & Business Media, 2012.
2. Liu, L.; Zhuang, Z.; Xie, T.; Wang, Y. G.; Li, J.; Peng, Q.; Li, Y. *J. Am. Chem. Soc.* **2009,** *131,* 16423–16429.
3. Huang, S.; Wang, D.; Li, C.; Wang, L.; Zhang, X.; Wan, Y.; Yang, P. *CrystEngComm* **2012,** *14,* 2235–2244.
4. Li, P.; Zhao, X.; Jia, C. J.; Sun, H.; Li, Y.; Sun, L.; Cheng, X.; Liu, L.; Fan, W. *Cryst. Growth Des.* **2012,** *12,* 5042–5050.
5. Peng, Z. A.; Peng, X. *J. Am. Chem. Soc.* **2001,** *123,* 1389–1395.
6. Deng, H.; Liu, C.; Yang, S.; Xiao, S.; Zhou, Z. K.; Wang, Q. Q. *Cryst. Growth Des.* **2008,** *8,* 4432–4439.
7. Siegfried, M. J.; Choi, K. S. *J. Am. Chem. Soc.* **2006,** *128,* 10356–10357.

8. Zhu, L.; Li, Q.; Liu, X.; Li, J.; Zhang, Y.; Meng, J.; Cao, X. *J. Phys. Chem. C* **2007**, *111*, 5898–5903.
9. Wang, M.; Huang, Q. L.; Hong, J. M.; Chen, X. T.; Xue, Z. L. *Cryst. Growth Des.* **2006**, *6*, 2169–2173.
10. Gu, J.; Zhu, Y.; Li, H.; Zhang, X.; Qian, Y. *J. Solid State Chem.* **2010**, *183*, 497–503.
11. Filankembo, A.; Giorgio, S.; Lisiecki, I.; Pileni, M. P. *J. Phys. Chem. B* **2003**, *107*, 7492–7500.
12. Wang, M.; Jiang, G.; Tang, Y.; Shi, Y. *CrystEngComm* **2013**, *15*, 1001–1006.
13. Li, C.; Quan, Z.; Yang, P.; Yang, J.; Lian, H.; Lin, J. *J. Mater. Chem.* **2008**, *18*, 1353–1361.
14. Li, C.; Yang, J.; Quan, Z.; Yang, P.; Kong, D.; Lin, J. *Chem. Mater.* **2007**, *19*, 4933–4942.
15. Kim, H. B.; Jang, D. J. *CrystEngComm* **2012**, *14*, 6946–6951.
16. Sayed, F. N.; Polshettiwar, V. *Sci. Rep.* **2015**, *5*, 9733.
17. Silver, J.; Martinez Rubio, M.; Ireland, T.; Fern, G.; Withnall, R. *J. Phys. Chem. B* **2001**, *105*, 948–953.
18. Na, H.; Woo, K.; Lim, K.; Jang, H. S. *Nanoscale* **2013**, *5*, 4242–4251.
19. Xu, Z.; Li, C.; Hou, Z.; Peng, C.; Lin, J. *CrystEngComm* **2011**, *13*, 474–482.
20. Rodriguez-Liviano, S.; Aparicio, F. J.; Rojas, T. C.; Hungría, A. B.; Chinchilla, L. E.; Ocaña, M. *Cryst. Growth Des.* **2011**, *12*, 635–645.
21. Xue, N.; Fan, X.; Wang, Z.; Wang, M. *J. Phys. Chem. Solids* **2008**, *69*, 1891–1896.
22. Liu, J.; Cano Torres, J. M.; Cascales, C.; Esteban Betegón, F.; Serrano, M. D.; Volkov, V.; Zaldo, C.; Rico, M.; Griebner, U.; Petrov, V. *Phys. Status Solidi A* **2005**, *202*, R29–R31.
23. Zhang, N.; Bu, W.; Xu, Y.; Jiang, D.; Shi, J. *J. Phys. Chem. C* **2007**, *111*, 5014–5019.
24. Karol, P. J. *The Periodic Table of the Elements: A Review of the Future, Elements Old and New: Discoveries, Developments, Challenges, and Environmental Implications*; American Chemical Society, 2017; pp 41–66.
25. Judd, B. *J. Chem. Phys.* **1966**, *44*, 839–840.
26. Mahalingam, V.; Vetrone, F.; Naccache, R.; Speghini, A.; Capobianco, J. A. *Adv. Mater.* **2009**, *21*, 4025–4028.
27. Maas, H.; Currao, A.; Calzaferri, G. *Angew. Chem. Int. Ed.* **2002**, *41*, 2495.
28. Stouwdam, J. W.; Hebbink, G. A.; Huskens, J.; van Veggel, F. C. J. M. *Chem. Mater.* **2003**, *15*, 4604.
29. Yanagida, S.; Hasegawa, Y.; Murakoshi, K.; Wada, Y.; Nakashima, N.; Yamanaka, T. *Coord. Chem. Rev.* **1998**, *171*, 461.
30. Andriessen, J.; van der Kolk, E.; Dorenbos, P. *Phys. Rev. B* **2007**, *76*, 075124.
31. Boyer, J. C.; Manseau, M. P.; Murray, J. I.; van Veggel, F. C. J. M. *Langmuir* **2009**, *26*, 1157.
32. Kar, A.; Patra, A. *Nanoscale* **2012**, *4*, 3608.
33. Paudel, H. P.; Zhong, L.; Bayat, K.; Baroughi, M. F.; Smith, S.; Lin, C.; Jiang, C.; Berry, M. T.; May, P. S. *J. Phys. Chem. C* **2011**, *115*, 19028.
34. Wegh, R. T.; Donker, H.; Oskam, K. D.; Meijerink, A. *Science* **1999**, *283*, 663–666.
35. Yang, J.; Li, C.; Quan, Z.; Zhang, C.; Yang, P.; Li, Y.; Yu, C.; Lin, J. *J. Phys. Chem. C* **2008**, *112*, 12777–12785.
36. Quici, S.; Cavazzini, M.; Marzanni, G.; Accorsi, G.; Armaroli, N.; Ventura, B.; Barigelletti, F. *Inorg. Chem.* **2005**, *44*, 529–537.
37. Stouwdam, J. W.; van Veggel, F. C. J. M. *Nano Lett.* **2002**, *2*, 733–737.

38. Boyer, J. C.; Cuccia, L. A.; Capobianco, J. A. *Nano Lett.* **2007**, *7*, 847–852.
39. Giri, N. K.; Singh, A. K.; Rai, S. B. *J. Appl. Phys.* **2007**, *101*, 033102.
40. Moore, E. G.; Samuel, A. P. S.; Raymond, K. N. *Acc. Chem. Res.* **2009**, *42*, 542–552.
41. Gangwar, P.; Pandey, M.; Sivakumar, S.; Pala, R. G. S.; Parthasarathy, G. *Cryst. Growth Des.* **2013**, *13*, 2344–2349.
42. Tanner, P. A. *Chem. Soc. Rev.* **2013**, *42*, 5090–5101.
43. Rastogi, C. K.; Saha, S.; Sivakumar, S.; Pala, R. G. S.; Kumar, J. *Phys. Chem. Chem. Phys.* **2015**, *17*, 4600–4608.
44. Stouwdam, J. W.; van Veggel, F. C. *ChemPhysChem* **2004**, *5*, 743–746.
45. Singh, S.; Rao, M. S. R. *Phys. Rev. B* **2009**, *80*, 045210.
46. Patra, A.; Friend, C. S.; Kapoor, R.; Prasad, P. N. *Appl. Phys. Lett.* **2003**, *83*, 284–286.
47. Kar, A.; Patra, A. *J. Phys. Chem. C* **2009**, *113*, 4375–4380.
48. Wang, F.; Liu, X. *Chem. Soc. Rev.* **2009**, *38*, 976–989.
49. Wang, L. P.; Chen, L. M. *Mater. Charact.* **2012**, *69*, 108–114.
50. Jia, C. J.; Sun, L. D.; Yan, Z. G.; Pang, Y. C.; Lü, S. Z.; Yan, C. H. *Eur. J. Inorg. Chem.* **2010**, *2010*, 2626–2635.
51. Fan, W.; Song, X.; Bu, Y.; Sun, S.; Zhao, X. *J. Phys. Chem. B* **2006**, *110*, 23247–23254.
52. Jia, C. J.; Sun, L. D.; You, L. P.; Jiang, X. C.; Luo, F.; Pang, Y. C.; Yan, C. H. *J. Phys. Chem. B* **2005**, *109*, 3284–3290.
53. Jia, C. J.; Sun, L. D.; Luo, F.; Jiang, X. C.; Wei, L. H.; Yan, C. H. *Appl. Phys. Lett.* **2004**, *84*, 5305–5307.
54. Shannon, R. *Acta Crystallogr. Sect. A* **1976**, *32*, 751–767.
55. Trojan-Piegza, J.; Zych, E.; Hölsä, J.; Niittykoski, J. *J. Phys. Chem. C* **2009**, *113*, 20493–20498.
56. Armelao, L.; Bottaro, G.; Pascolini, M.; Sessolo, M.; Tondello, E.; Bettinelli, M.; Speghini, A. *J. Phys. Chem. C* **2008**, *112*, 4049–4054.
57. Julián, B.; Corberán, R.; Cordoncillo, E.; Escribano, P.; Viana, B.; Sanchez, C. *Nanotechnology* **2005**, *16*, 2707.
58. Jeyaraman, J.; Shukla, A.; Sivakumar, S. *ACS Biomater. Sci. Eng.* **2016**, *2*, 1330–1340.
59. Yin, W.; Zhou, L.; Gu, Z.; Tian, G.; Jin, S.; Yan, L.; Liu, X.; Xing, G.; Ren, W.; Liu, F.; Pan, Z.; Zhao, Y. *J. Mater. Chem.* **2012**, *22*, 6974–6981.
60. Park, C. S.; Tae, H. S.; Jung, E. Y.; Seo, J. H.; Shin, B. J. *IEEE Trans. Plasma Sci.* **2010**, *38*, 2439–2444.
61. Uchino, T.; Okutsu, D.; Katayama, R.; Sawai, S. *Phys. Rev. B* **2009**, *79*, 165107.
62. Mai, H. X.; Zhang, Y. W.; Si, R.; Yan, Z. G.; Sun, L.; You, L. P.; Yan, C. H.; *J. Am. Chem. Soc.* **2006**, *128*, 6426–6436.
63. Wang, L.; Li, Y. *Nano Lett.* **2006**, *6*, 1645–1649.
64. Xie, R. -J.; Hirosaki, N.; Li, Y.; Takeda, T. *Materials* **2010**, *3*, 3777–3793.
65. Parchur, A. K.; Ningthoujam, R. S. *RSC Adv.* **2012**, *2*, 10859–10868.
66. Singh, S.; Tripathi, A.; Kumar Rastogi, C.; Sivakumar, S. *RSC Adv.* **2012**, *2*, 12231–12236.
67. Li, X.; Wang, R.; Zhang, F.; Zhou, L.; Shen, D.; Yao, C.; Zhao, D. *Sci. Rep.* **2013**, *3*, 3536.
68. Lian, J.; Qin, H.; Liang, P.; Liu, F. *Solid State Sci.* **2015**, *48*, 147–154.
69. Liu, X.; Li, L.; Noh, H. M.; Jeong, J. H.; Jang, K.; Shin, D. S. *RSC Adv.* **2015**, *5*, 9441–9454.

70. Parchur, A. K.; Ningthoujam, R. S. *RSC Adv.* **2012**, *2*, 10854–10858.
71. Lin, J.; Wang, Q. *Chem. Eng. J.* **2014**, *250*, 190–197.
72. Mi, R.; Chen, J.; Liu, Y. G.; Fang, M.; Mei, L.; Huang, Z.; Wang, B.; Zhaob, C. *RSC Adv.* **2016**, *6*, 28887–28894.
73. Qin, X.; Yokomori, T.; Ju, Y. *Appl. Phys. Lett.* **2007**, *90*, 073104.
74. Liu, C.; Chen, D. *J. Mater. Chem.* **2007**, *17*, 3875–3880.
75. Ehlert, O.; Thomann, R.; Darbandi, M.; Nann, T. *ACS Nano* **2008**, *2*, 120–124.
76. Heer, K. K. S.; Güdel, H. U.; Haase, M. *Adv. Mater.* **2004**, *16*, 2102.
77. Yi, G. S.; Chow, G. M. *Adv. Funct. Mater.* **2006**, *16*, 2324.
78. Liu, C.; Chen, D. *J. Mater. Chem.* **2007**, *17*, 3875–3880.
79. Stouwdam, J. W.; van Veggel, F. C. J. M. *Langmuir* **2004**, *20*, 11763–11771.
80. Shan, J.; Uddi, M.; Wei, R.; Yao, N.; Ju, Y. *J. Phys. Chem. C* **2010**, *114*, 2452–2461.
81. Mai, H. X.; Zhang, Y. W.; Sun, L. D.; Yan, C. H. *J. Phys. Chem. C* **2007**, *111*, 13730–13739.
82. Naidu, B. S.; Vishwanadh, B.; Sudarsan, V.; Vatsa, R. K. *Dalton Trans.* **2012**, *41*, 3194–3203.
83. Singh, L. R.; Ningthoujam, R.; Sudarsan, V.; Srivastava, I.; Singh, S. D.; Dey, G.; Kulshreshtha, S. *Nanotechnology* **2008**, *19*, 055201.
84. Liu, Y.; Tu, D.; Zhu, H.; Chen, X. *Chem. Soc. Rev.* **2013**, *42*, 6924–6958.
85. Qin, X.; Liu, X.; Huang, W.; Bettinelli, M.; Liu, X. *Chem. Rev.* **2017**, *117*, 4488–4527.
86. Znaidi, L. *Mater. Sci. Eng. B* **2010**, *174*, 18–30.
87. Sekar, S.; Antony, A. G.; Vijayan, V.; Marappan, L.; Baskar, S. *Int. J. Mech. Eng. Technol.* **2019**, 785–790.
88. Sun, X.; Zhang, Y. W.; Du, Y. P.; Yan, Z. G.; Si, R.; You, L. P.; Yan, C. H. *Chem. Eur. J.* **2007**, *13*, 2320–2332.
89. Zhang, Y. -W.; Sun, X.; Si, R.; You, L. -P.; Yan, C. -H. *J. Am. Chem. Soc.* **2005**, *127*, 3260–3261.
90. Du, Y. P.; Zhang, Y. W.; Sun, L. D.; Yan, C. H. *J. Phys. Chem. C* **2008**, *112*, 405–415.
91. Du, Y. P.; Zhang, Y. W.; Yan, Z. G.; Sun, L. D.; Yan, C. H. *J. Am. Chem. Soc.* **2009**, *131*, 16364–16365.
92. Du, Y. P.; Zhang, Y. W.; Sun, L. D.; Yan, C. H. *J. Am. Chem. Soc.* **2009**, *131*, 3162–3163.
93. Du, Y. P.; Zhang, Y. W.; Sun, L. D.; Yan, C. H. *Dalton Trans.* **2009**, 8574–8581.
94. Zhang, Y. W.; Sun, X.; Si, R.; You, L. P.; Yan, C. H. *J. Am. Chem. Soc.* **2005**, *127*, 3260–3261.
95. Boyer, J. C.; Vetrone, F.; Cuccia, L. A.; Capobianco, J. A. *J. Am. Chem. Soc.* **2006**, *128*, 7444–7445.
96. Yi, G. S.; Chow, G. M. *Adv. Funct. Mater.* **2006**, *16*, 2324–2329.
97. Jingning Shan, X. Q.; Yao, N.; Ju, Y. *Nanotechnology* **2007**, *18*, 1–7.
98. Wang, J.; Wang, F.; Xu, J.; Wang, Y.; Liu, Y.; Chen, X.; Chen, H.; Liu, X. *C. R. Chim.* **2010**, *13*, 731–736.
99. Si, R.; Zhang, Y. W.; Zhou, H. P.; Sun, L. D.; Yan, C. H. *Chem. Mater.* **2007**, *19*, 18–27.
100. Heer, S.; Lehmann, O.; Haase, M.; Güdel, H. -U. *Angew. Chem. Int. Ed.* **2003**, *42*, 3179–3182.
101. Sun, J.; Mi, X.; Lei, L.; Pan, X.; Chen, S.; Wang, Z.; Bai, Z.; Zhang, X. *CrystEngComm* **2015**, *17*, 7888–7895.
102. Lin, C. C.; Lin, K. M.; Li, Y. Y. *J. Lumin.* **2007**, *126*, 795–799.

103. Liu, C.; Wang, H.; Zhang, X.; Chen, D. *J. Mater. Chem.* **2009,** *19*, 489–496.
104. Shan, J.; Ju, Y. *Appl. Phys. Lett.* **2007,** *91*, 123103.
105. Rastogi, C. K.; Saha, S.; Kusuma, V.; Pala, R. G. S.; Kumar, J.; Sivakumar, S. *Cryst. Grow Des.* **2019,** *19*, 3945–3954.
106. Qiu, H.; Chen, G.; Fan, R.; Cheng, C.; Hao, S.; Chen, D.; Yang, C. *Chem. Commun.* **2011,** *47*, 9648–9650.
107. Zhang, Y.; Zhang, H.; Xu, Y.; Wang, Y. *J. Mater. Chem.* **2003,** *13*, 2261–2265.
108. Li, C.; Yang, J.; Quan, Z.; Yang, P.; Kong, D.; Lin, J. *Chem. Mater.* **2007,** *19*, 4933–4942.
109. Rastogi, C. K.; Sharma, S. K.; Patel, A.; Parthasarathy, G.; Pala, R. G. S.; Kumar, J.; Sivakumar, S. *J. Phys. Chem. C* **2017,** *121*, 16501–16512.
110. Dai, Q.; Foley, M. E.; Breshike, C. J.; Lita, A.; Strouse, G. F. *J. Am. Chem. Soc.* **2011,** *133*, 15475–15486.
111. Hou, Z.; Li, C.; Ma, P. A.; Cheng, Z.; Li, X.; Zhang, X.; Dai, Y.; Yang, D.; Lian, H.; Lin, J. *Adv. Funct. Mater.* **2012,** *22*, 2713–2722.
112. Jeyaraman, J.; Malecka, A.; Billimoria, P.; Shukla, A.; Marandi, B.; Patel, P. M.; Jackson, A. M.; Sivakumar, S. *J. Mater. Chem. B* **2017,** *5*, 5251–5258.
113. Ganguli, S.; Hazra, C.; Chatti, M.; Samanta, T.; Mahalingam, V. *Langmuir* **2016,** *32*, 247–253.
114. Kumar, P.; Gupta, B. K. *RSC Adv.* **2015,** *5*, 24729–24736.
115. Lin, C. C.; Zheng, Y. S.; Chen, H. Y.; Ruan, C. H.; Xiao, G. W.; Liu, R. S. *J. Electrochem. Soc.* **2010,** *157*, H900–H903.
116. Zhu, H.; Lin, C. C.; Luo, W.; Shu, S.; Liu, Z.; Liu, Y.; Kong, J.; Ma, E.; Cao, Y.; Liu, R. -S.; Chen, X. *Nat. Commun.* **2014,** *5*, 4312.
117. Bünzli, J. C. G. *Chem. Rev.* **2010,** *110*, 2729–2755.
118. Dong, H.; Du, S. R.; Zheng, X. Y.; Lyu, G. M.; Sun, L. D.; Li, L. D.; Zhang, P. Z.; Zhang, C.; Yan, C. H. *Chem. Rev.* **2015,** *115*, 10725–10815.
119. Zhou, J.; Liu, Q.; Feng, W.; Sun, Y.; Li, F. *Chem. Rev.* **2015,** *115*, 395–465.
120. van der Ende, B. M.; Aarts, L.; Meijerink, A. *Phys. Chem. Chem. Phys.* **2009,** *11*, 11081–11095.

CHAPTER 5

Quantitative Estimation of Heavy Metals (As, Hg, Cd, and Pb) in Human Blood Samples Collected from Different Areas of Rajasthan

NIDHI GAUR[1*], GINNI KUMAWAT[2], ROMILA KARNAWAT[3], I. K. SHARMA[1], and P. S. VERMA[1]

[1]*Department of Chemistry, University of Rajasthan, Jaipur, Rajasthan, India*

[2]*Forensic Science Laboratory, Jaipur, Rajasthan, India*

[3]*Govt. SCRS College, Sawaimadhopur, Rajasthan, India*

Corresponding author. E-mail: gaurnidhi1985@gmail.com

ABSTRACT

Rapid urbanization and industrialization in many developing countries have given rise to contamination of natural resources. Globally everyone is potentially vulnerable to the toxic effects of heavy metals. Many studies carried out throughout India reveal an increase in concentration of heavy metals in our surroundings. The aim of the present study was to determine the concentration of heavy metals (As, Hg, Cd, and Pb) in human blood samples collected from Jaipur, Ajmer, Kota, Jodhpur, Udaipur, and Bharatpur regions of Rajasthan. The quantitative estimation of heavy metals was carried out by atomic absorption spectroscopy. The concentrations of these heavy metals were found greater than the permissible limit in some of the blood samples, which is an alarming situation. Statistical analysis of obtained data also discloses a nonsignificant difference in the concentration of these heavy metals in different areas, which means that all are equally vulnerable to heavy metal toxicity.

5.1   INTRODUCTION

Environmental degradation due to uncontrolled anthropogenic activities has become a major societal issue. The fast expansion of urban, agricultural, and industrial activities spurred by rapid population growth and the change in consumer habits has produced vast amounts of solid waste.[1] Globally everyone is potentially vulnerable to the toxic effects of heavy metals. Many toxic heavy metals are ubiquitous in our environment. Researches in the field of environmental medicine reveal the detrimental effects of heavy metals on the functioning of heart, immune system, nervous system, etc.[2]

Heavy metal toxicity has received special attention globally due to neurotoxin, carcinogenic, and several other impacts arising from their consumption even at lower contents. Heavy metals have impacted the ecosystem through discharges as effluents, dust, and leachate. These elements exhibit varying environmental behavior and toxicity to aquatic organism and man. Although many metals are essential, all metals are toxic at elevated concentrations, because they form free radicals thereby causing oxidative stress and can replace essential metals in enzymes disrupting their function.[3] Many trace elements are biologically beneficial at very low concentrations but become toxic or otherwise detrimental to the health of organisms and plants at low-to-moderate concentrations.

Various anthropogenic activities without taking any safety measures have caused the problem of heavy metal pollution in the soils. Heavy metals are a large group of elements with higher density generally greater than 5 g/cm^3.[4] These elements are important both industrially and biologically. Heavy metals occur naturally in the earth's crust and surface soils in varying concentrations.[5] Natural processes like weathering and erosion remove small amounts of metals from the bedrocks and allow them to circulate in water and air. Heavy metals are very harmful in reference to their nonbiodegradable nature, long biological half-lives, and their potential to accumulate in different body parts.[6] There are more than 20 heavy metals, but 4 are of particular concern to human health and the environment, lead (Pb), cadmium (Cd), mercury (Hg), and arsenic (As).[7]

Although several adverse health effects of heavy metals have been known for a long time, exposure to these metals is continuing and even increasing in some parts of the world. Thus, the control of heavy metal dumplings and the removal of toxic heavy metals from waters have become a challenge for the 21st century.

In developing countries like India, heavy metal poisoning remains a serious problem. Water Quality Assessment Authority has been constituted by the Government of India to monitor the issues regarding the quality of water in Indian rivers.[8] Trace and toxic metals in the river water have large bearing on the health of human being, aquatic life, and ecology. In a report submitted by Central Water Commission in 2014, it has been observed that the concentration of some of the heavy metals in Indian river system is near or more than permissible limit that is a matter of concern.

West Bengal, Bihar, UP, Assam, and Chhattisgarh are the major states affected by arsenic contamination of water, West Bengal being by far the worst affected.[9] Arsenic contamination in groundwater in the Ganga–Brahmaputra plains and Padma–Meghna plains in Bangladesh and its consequences have been reported to be a serious threat to human health.[10] Despite several precautionary measures, the spread of arsenic contamination in groundwater has been continued to grow and now it has been reported in other regions as well.[11] Groundwater-led contamination is also a serious issue. Lead is considered to be one of the most serious environmental poisons.[12]

The measurement of metal levels is helpful in ascertaining risk to human health as well as in the assessment of environmental quality.[13] Many reports indicated heavy metals in human blood and attributed the presence of these heavy metals due to exposure of human beings to environmental pollution. Blood and urine samples reflect the amount of metals circulating at the time of sampling and do not represent the cumulative degree of exposure. Tissue biopsies for elemental analysis can aid in identifying the accumulation of metals from chronic exposure.[14] Human biomonitoring is a method of accurately measuring contact and absorption of toxic chemicals from the environment.[15] It is the direct measurement of an individual's environment.

The present study was planned to estimate biomonitoring of selected heavy metals (As, Hg, Cd, and Pb) in blood samples of some volunteers of selected areas who were either directly or indirectly exposed to heavy metals. The objective of this study is to analyze heavy metal contamination in blood samples of human beings belonging to different age groups of various regions in Rajasthan employing qualitative and quantitative analysis. An attempt has been made to determine the presence of heavy metal ions in human blood samples and quantification has been done by graphite furnace atomic absorption spectroscopy (AAS).

5.2 MATERIAL AND METHOD

5.2.1 COLLECTION OF BLOOD SAMPLES

The blood samples were collected randomly from the general population from different regions of Rajasthan, namely, Ajmer region, Jaipur region, Kota region, Jodhpur region, Udaipur region, and Bharatpur region. These samples were kept in a plastic container and stored at 5°C until analysis.

5.2.2 DIGESTION OF BLOOD SAMPLES

The blood samples were digested in the presence of conc. HNO_3 as per the procedure discussed earlier. The extracts of human blood samples obtained from the abovementioned procedure were further subjected for quantitative analysis of selected heavy metals (As, Hg, Cd, and Pb) using AAS.

5.2.3 INSTRUMENT

Atomic absorption spectrophotometer (EC Electronics Corporation of India Limited, AAS Element AS AAS4141) equipped with a deuterium lamp for background correction has been used for determination of heavy metal ions in human blood samples digested by conc. HNO_3.
The details of instrumental conditions are given in Table 5.1.

TABLE 5.1   Instrumental Conditions.

Element	Current (mA)	Slit width (nm)	λmax (nm)	Flame color	Flame type	AAS technique
Cd	3.5	0.5	228.8	Blue	Air/C_2H_2	Flame
Pb	10	1.0	217	Blue	Air/C_2H_2	Flame
As	EDL (expected detectable limit)	1	193.7	Blue	Air/C_2H_2	Hydride generation
Hg	EDL	0.5	253.6	–	–	Cold vapor

The concentrations of selected heavy metals (As, Hg, Cd, and Pb) in collected human blood samples from various regions of Rajasthan were determined by AAS and the obtained values are given in Tables 5.2–5.7.

5.3 OBSERVATION TABLE

TABLE 5.2 Concentrations of Selected Heavy Metals (As, Pb, Cd, and Hg) in Human Blood Samples Collected from Ajmer Region.

S. No.	Sample	Concentration of arsenic (in ppm)	Concentration of lead (in ppm)	Concentration of cadmium (in ppm)	Concentration of mercury (in ppm)
1	Sample no. 1	2.09	4.117	0.558	0.001
2	Sample no. 2	2.15	10.332	0.413	0.02
3	Sample no. 3	2.17	6.332	0.197	0.05
4	Sample no. 4	3.38	11.424	0.812	0.02
5	Sample no. 5	1.06	8.886	0.452	0.01
6	Sample no. 6	2.17	5.445	0.632	0.06
7	Sample no. 7	1.98	11.232	0.548	0.006
8	Sample no. 8	2.23	7.444	0.221	0.002
9	Sample no. 9	2.37	10.044	0.738	0.12
10	Sample no. 10	0.98	9.887	0.339	0.08

TABLE 5.3 Concentrations of Selected Heavy Metals (As, Pb, Cd, and Hg) in Human Blood Samples Collected from Jaipur Region.

S. No.	Sample	Concentration of arsenic (in ppm)	Concentration of lead (in ppm)	Concentration of cadmium (in ppm)	Concentration of mercury (in ppm)
11	Sample no. 11	1.79	2.178	0.334	0.003
12	Sample no. 12	1.02	5.36	0.194	0.16
13	Sample no. 13	2.07	3.334	0.102	0.02
14	Sample no. 14	2.79	9.818	0.644	0.02
15	Sample no. 15	0.89	3.225	0.167	0.005
16	Sample no. 16	2.36	3.339	0.245	0.01
17	Sample no. 17	1.17	1.878	0.331	0.003
18	Sample no. 18	2.46	5.287	0.712	0.04
19	Sample no. 19	0.994	4.286	0.114	0.003
20	Sample no. 20	3.89	5.874	0.708	0.004

TABLE 5.4   Concentrations of Selected Heavy Metals (As, Pb, Cd, and Hg) in Human Blood Samples Collected from Kota Region.

S. No.	Sample	Concentration of arsenic (in ppm)	Concentration of lead (in ppm)	Concentration of cadmium (in ppm)	Concentration of mercury (in ppm)
21	Sample no. 21	2.23	6.667	0.103	0.001
22	Sample no. 22	3.48	2.339	0.508	0.003
23	Sample no. 23	1.16	11.56	0.612	0.02
24	Sample no. 24	1.38	3.885	0.78	0.01
25	Sample no. 25	3.78	6.72	0.809	0.005
26	Sample no. 26	2.65	10.88	0.110	0.01
27	Sample no. 27	2.85	9.82	0.655	0.001
28	Sample no. 28	1.17	1.99	0.339	0.03
29	Sample no. 29	2.24	2.348	0.113	0.003
30	Sample no. 30	2.92	4.78	0.789	0.08

TABLE 5.5   Concentrations of Selected Heavy Metals (As, Pb, Cd, and Hg) in Human Blood Samples Collected from Jodhpur Region.

S. No.	Sample	Concentration of arsenic (in ppm)	Concentration of lead (in ppm)	Concentration of cadmium (in ppm)	Concentration of mercury (in ppm)
31	Sample no. 31	3.45	11.874	0.665	0.05
32	Sample no. 32	2.69	4.79	0.125	0.008
33	Sample no. 33	1.52	2.879	0.657	0.09
34	Sample no. 34	3.33	6.98	0.327	0.10
35	Sample no. 35	3.59	7.56	0.112	0.02
36	Sample no. 36	2.72	5.74	0.123	0.01
37	Sample no. 37	2.17	9.87	0.135	0.001
38	Sample no. 38	1.89	11.768	0.165	0.07
39	Sample no. 39	2.92	8.77	0.768	0.004
40	Sample no. 40	3.48	3.65	0.558	0.03

TABLE 5.6 Concentrations of Selected Heavy Metals (As, Pb, Cd, and Hg) in Human Blood Samples Collected from Udaipur Region.

S. No.	Sample	Concentration of arsenic (in ppm)	Concentration of lead (in ppm)	Concentration of cadmium (in ppm)	Concentration of mercury (in ppm)
41	Sample no. 41	2.25	5.79	0.456	0.006
42	Sample no. 42	2.58	1.998	0.345	0.13
43	Sample no. 43	1.19	3.79	0.595	0.06
44	Sample no. 44	2.77	8.54	0.110	0.12
45	Sample no. 45	3.79	5.89	0.459	0.04
46	Sample no. 46	2.35	9.78	0.767	0.003
47	Sample no. 47	2.67	10.57	0.556	0.12
48	Sample no. 48	3.39	2.976	0.782	0.01
49	Sample no. 49	1.87	5.889	0.119	0.004
50	Sample no. 50	1.85	7.556	0.175	0.005

TABLE 5.7 Concentrations of Selected Heavy Metals (As, Pb, Cd, and Hg) in Human Blood Samples Collected from Bharatpur Region.

S. No.	Sample	Concentration of arsenic (in ppm)	Concentration of lead (in ppm)	Concentration of cadmium (in ppm)	Concentration of mercury (in ppm)
51	Sample no. 51	2.99	4.78	0.187	0.08
52	Sample no. 52	3.48	3.97	0.112	0.01
53	Sample no. 53	3.84	6.79	0.805	0.002
54	Sample no. 54	1.14	3.98	0.812	0.14
55	Sample no. 55	2.09	11.58	0.768	0.07
56	Sample no. 56	2.98	9.87	0.805	0.01
57	Sample no. 57	1.72	5.689	0.506	0.006
58	Sample no. 58	1.92	4.08	0.406	0.12
59	Sample no. 59	3.67	6.98	0.782	0.02
60	Sample no. 60	2.89	10.99	0.545	0.003

The graphical representation of concentration of selected heavy metals in analyzed samples is shown in the following figures (Figs. 5.1–5.6):

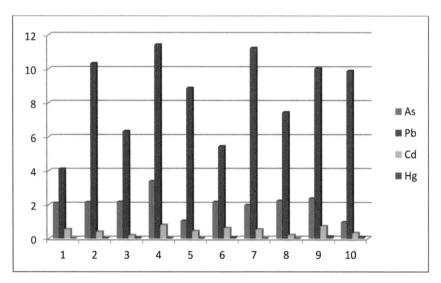

FIGURE 5.1 Concentrations of selected heavy metals in human blood samples collected from Ajmer region.

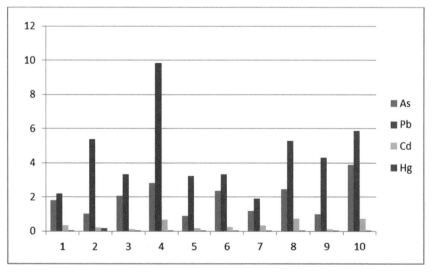

FIGURE 5.2 Concentrations of selected heavy metals in human blood samples collected from Jaipur region.

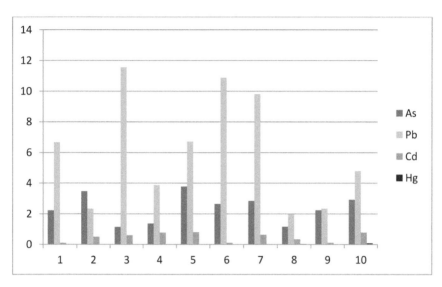

FIGURE 5.3 Concentrations of selected heavy metals in human blood samples collected from Kota region.

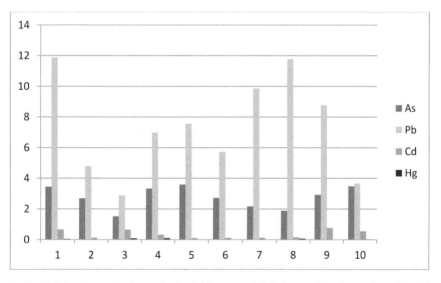

FIGURE 5.4 Concentrations of selected heavy metals in human blood samples collected from Jodhpur region.

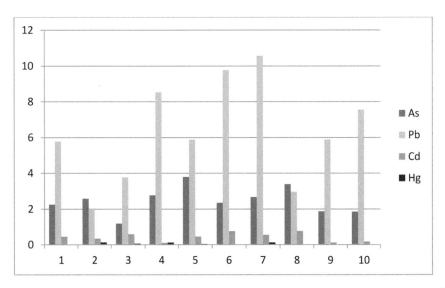

FIGURE 5.5   Concentrations of selected heavy metals in human blood samples collected from Udaipur region.

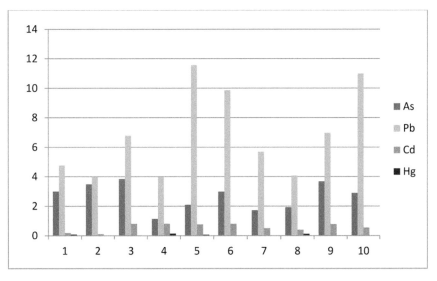

FIGURE 5.6   Concentrations of selected heavy metals in human blood samples collected from Bharatpur region.

5.4 STATISTICAL ANALYSIS

The statistical analysis of concentrations of selected heavy metals (As, Hg, Cd, and Pb) determined by AAS in human blood samples collected from different areas of Rajasthan was carried out by SPSS (Statistical Package for the Social Sciences) software. The selected heavy metals are taken as four descriptives and obtained results are discussed next.

5.5 MEAN VALUES AND STANDARD DEVIATION

TABLE 5.8 Mean Values and Standard Deviation.

	N	**Mean**	**Standard deviation**
Arsenic	60	2.3844	0.85358
Lead	60	6.6013	3.08509
Cadmium	60	0.4501	0.25638
Mercury	60	0.0334	0.04033

Here, *N* is a number of samples.

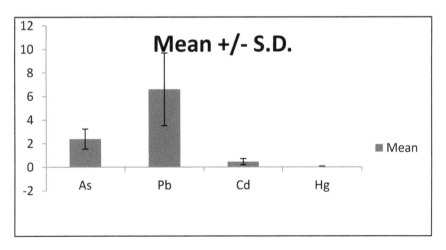

FIGURE 5.7 Graphical representation of mean of concentration of heavy metals (As, Pb, Cd, and Hg) in collected blood samples.

5.5.1 ONE-WAY ANOVA FOR ARSENIC

TABLE 5.9 One-way ANOVA for Concentration of Arsenic in Human Blood Samples Collected from Different Regions of Rajasthan.

S. No.	Regions	N	Mean	Standard deviation	Standard error	95% confidence interval for mean	
						Lower bound	Upper bound
1.	Ajmer	10	2.0580	0.67275	0.21274	1.5767	2.5393
2.	Jaipur	10	1.9434	0.96884	0.30637	1.2503	2.6365
3.	Jodhpur	10	2.3860	0.92794	0.29344	1.7222	3.0498
4.	Kota	10	2.7760	0.72114	0.22805	2.2601	3.2919
5.	Udaipur	10	2.4710	0.75840	0.23983	1.9285	3.0135
6.	Bharatpur	10	2.6720	0.90763	0.28702	2.0227	3.3213
	Total	**60**	2.3844	0.85358	0.11020	2.1639	2.6049

ANOVA

Arsenic

	Sum of squares	df	Mean square	F	Significance
Between groups	5.446	5	1.089		
Within groups	37.542	54	0.695		
Total	42.988	59		1.567	0.185

Here, N is a number of blood samples collected from different regions.

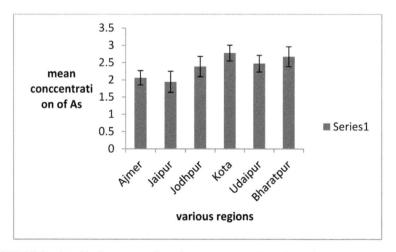

FIGURE 5.8 Graphical representation of mean concentration of arsenic in various regions.

5.5.2 ONE-WAY ANOVA FOR Pb

TABLE 5.10 One-way ANOVA for Concentration of Lead in Human Blood Samples Collected from Different Regions of Rajasthan.

S. No.	Regions	N	Mean	Standard deviation	Standard error	95% confidence interval for mean	
						Lower bound	Upper bound
1.	Ajmer	10	8.5143	2.54260	0.80404	6.6954	10.3332
2.	Jaipur	10	4.4579	2.31314	0.73148	2.8032	6.1126
3.	Jodhpur	10	6.0989	3.63574	1.14972	3.4980	8.6998
4.	Kota	10	7.3881	3.18580	1.00744	5.1091	9.6671
5.	Udaipur	10	6.2779	2.85554	0.90300	4.2352	8.3206
6.	Bharatpur	10	6.8709	2.95308	0.93385	4.7584	8.9834
	Total	**60**	6.6013	3.08509	0.39828	5.8044	7.3983

ANOVA

Lead

	Sum of squares	df	Mean square	F	Significance
Between groups	93.025	5	18.605		
Within groups	468.524	54	8.676		
Total	561.549	59		2.144	0.074

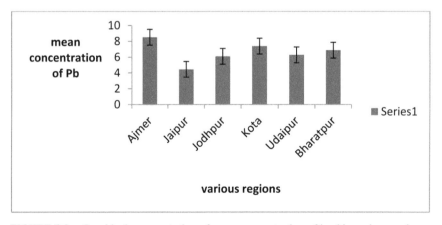

FIGURE 5.9 Graphical representation of mean concentration of lead in various regions.

5.5.3   ONE-WAY ANOVA FOR CONCENTRATION OF CADMIUM

TABLE 5.11   One-way ANOVA for Concentration of Cadmium in Human Blood Samples Collected from Different Regions of Rajasthan.

S. No.	Regions	*N*	Mean	Standard deviation	Standard error	95% confidence interval for mean	
						Lower bound	Upper bound
1.	Ajmer	10	0.4910	0.20594	0.06513	0.3437	0.6383
2.	Jaipur	10	0.3551	0.24314	0.07689	0.1812	0.5290
3.	Jodhpur	10	0.4818	0.29366	0.09286	0.2717	0.6919
4.	Kota	10	0.3635	0.26862	0.08495	0.1713	0.5557
5.	Udaipur	10	0.4364	0.24789	0.07839	0.2591	0.6137
6.	Bharatpur	10	0.5728	0.26691	0.08440	0.3819	0.7637
	Total	**60**	0.4501	0.25638	0.03310	0.3839	0.5163

ANOVA

Cadmium

	Sum of squares	df	Mean square	*F*	Significance
Between groups	0.344	5	0.069		
Within groups	3.534	54	0.065		
Total	3.878	59		1.053	0.397

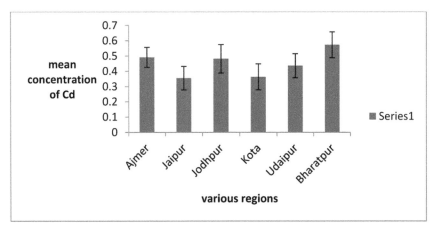

FIGURE 5.10   Graphical representation of mean concentration of cadmium in various regions.

5.5.4 ONE-WAY ANOVA FOR CONCENTRATION OF MERCURY

TABLE 5.12 One-way ANOVA for Concentration of Mercury in Human Blood Samples Collected from Different Regions of Rajasthan.

S. No.	Regions	N	Mean	Standard deviation	Standard error	95% confidence interval for mean	
						Lower bound	Upper bound
1.	Ajmer	10	0.0369	0.03976	0.01257	0.0085	0.0653
2.	Jaipur	10	0.0128	0.01215	0.00384	0.0041	0.0215
3.	Jodhpur	10	0.0163	0.02424	0.00767	0.0010	0.0336
4.	Kota	10	0.0383	0.03699	0.01170	0.0118	0.0648
5.	Udaipur	10	0.0498	0.05404	0.01709	0.0111	0.0885
6.	Bharatpur	10	0.0461	0.05238	0.01657	0.0086	0.0836
	Total	60	0.0334	0.04033	0.00521	0.0229	0.0438

ANOVA

Mercury

	Sum of squares	df	Mean square	F	Significance
Between groups	0.012	5	0.002		
Within groups	0.084	54	0.002		
Total	0.096	59		1.519	0.199

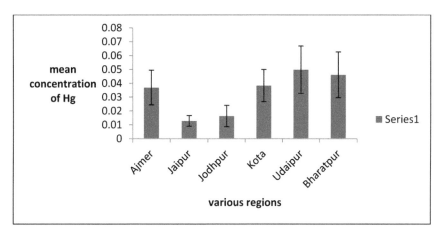

FIGURE 5.11 Graphical representation of mean concentration of mercury in various regions.

One-way ANOVA for the concentration of selected heavy metals (As, Pb, Cd, and Hg) in blood samples collected from different areas of Rajasthan was carried out. The p values for the concentration of As, Pb, Cd, and Hg from ANOVA were found to be 0.185, 0.074, 0.397, and 0.199. As p values for concentration of these heavy metals are greater than 0.05, it is clear that there are no statistically significant differences between the groups. The graphical representations of one-way ANOVA results are given in Figures 5.8–5.11.

5.6   RESULTS AND DISCUSSION

Some of the organizations concerned with human health like WHO, ATSDR (Agency for Toxic Substances and Disease Registry) have determined the lower limits of concentrations of these toxic heavy metals in human blood. According to ATSDR, the levels of arsenic in blood of unexposed individuals should be less than 1 ppm.[16] The geometric blood level of cadmium in the general population should be 0.315 ppm.[17] Similarly, geometric mean of lead levels in blood for children (1–5 years) should be 1.9 ppm and for adults (20–59 years) was 1.5 ppm[18] Whole blood mercury levels are usually <10 µg/L (ppb) in unexposed individuals.[19]

It is apparent from the analysis of observed data that concentration of arsenic in the blood samples collected randomly from different regions of Rajasthan is somewhat high. It may be due to occupational and nonoccupational exposure to arsenic. It has been recognized from ancient times that inorganic arsenic can act as poison and can result in death.[20] The US Environmental Protection Agency (EPA) has classified inorganic arsenic as a "human carcinogen" (maximum contaminant level 0.01 ppm).[21] Exposure mainly occurs through drinking groundwater contaminated with inorganic arsenic salts, and from food prepared or crops irrigated using high-arsenic water sources. The health effects can take years to manifest, depending on the level of arsenic in the drinking water, volume of intake, and nutritional status of individuals. In India, West Bengal is an arsenic endemic state where arsenic contamination of groundwater is above WHO permissible limit.[22] In some of the samples, concentration of arsenic has been found above the permissible limit, which is a matter of concern. Long-term intake of arsenic can lead to arsenic poisoning or arsenicosis with cancer of skin, bladder, kidney, or lung.

The concentration of lead is found to be appreciably high in the collected samples as is evident from the above tables and also can be seen in the following graphs. In many of the samples, concentration of Pb has been observed above permissible limit. It may be ascribed due to a constant raise in the level of pollution in the atmosphere. The major sources of lead contamination are automobiles and industrial untreated wastewater. Besides it, young children may also get exposed to lead-containing toys. According to a WHO study, lead exposure is predicted to account for 143,000 deaths per year with the highest burden in developing regions.[23] Young children are particularly susceptible to the toxic effects of lead and can suffer reflective and permanent adverse health effects, particularly affecting the development of the brain and nervous system.[24] Lead exposure is estimated to contribute to 600,000 new cases of children with intellectual disabilities every year.[25] Another study pointed out that Bangalore leads to lead toxicity in India that can be connected with city's phenomenal growth of vehicular traffic.[26]

The concentration of Cd is found to be more than permissible limit in some of the blood samples. The exposure of general population for Cd may be through food, cigarette smoking, drinking water, and air. Some occupations such as alloy production, battery production, pigment production, and plastic production have higher potential for cadmium exposure. Some recent studies reported that the level of cadmium was found above permissible limit in groundwater samples collected from some areas of Rajasthan.[27] A case study reported the risk of cadmium toxicity and interstitial pneumonitis even after a single brief exposure to cadmium fumes.[28] For the general nonsmoking population in most countries, food is the most important source of cadmium exposure. Cadmium is present in most foodstuffs, but concentrations differ greatly, and individual intake also varies considerably due to differences in dietary habits.[29]

It is noteworthy that the concentration of mercury has not been observed significant in most of the samples. Almost all the samples have been found to contain Hg below the permissible limit (1 ppm). In general, the information about distribution of mercury in blood system and hair mercury levels in Indian population is not easy to acquire. Mercury has its origin from natural as well as anthropogenic sources. The primary sources of human exposure to mercury are consumption of contaminated fish, occupational exposure, dental amalgam, vaccines, and domestic use.[30] In the last 100 years, 70% mercury is released in the environment from human activities as reported in the literature.[31] The global release of mercury to the environment is growing dramatically and has become a global threat.

5.7 CONCLUSION

In the present study, the concentrations of four selected heavy metal (As, Hg, Pb, and Cd) in human blood samples collected from different areas of Rajasthan were determined. The enhanced levels of concentrations of these metals especially have drawn public attention resulting from food safety issues and potential health risk. They are regarded as the most important contaminants in our environment.

Rajasthan has not been assumed a state, which is facing a threat of heavy metal contamination but the results of the present study suggest that the rising levels of these metals are alarming and a matter of concern. The enhanced levels of these heavy metals can disturb human body metabolism to some extent. The blood lead levels in the study areas are found to be higher than allowable limits for lead in blood samples. Lead is a heavy metal of wide occupational and environmental concern. Lead is accumulative poison. Accumulation of lead in the nervous system, bones, liver, pancreas, teeth, gums, and blood disturbs body function.

The results obtained from the present study point out that the concentration of mercury was observed to be comparatively lower than other heavy metals, which denotes that the general population is not so much exposed to mercury till now. The concentration of arsenic was also found to be above permissible limit in some samples. The presence of arsenic in a small quantity in water or food may affect the human health drastically and some time may lead to death. Somewhat lower doses produce subacute effects in the respiratory, gastrointestinal, cardiovascular, and nervous systems. When the concentration of cadmium was tested in collected blood samples, it was found greater than permissible limit in some of the samples. Tissue cadmium concentrations increase with age and kidney and liver act as cadmium stores. Due to this age-related accumulation of cadmium in body, only small part of cadmium is excreted through urine. Cadmium is carcinogenic and can cause DNA degradation.

From the present study, it may be concluded that heavy metal contamination is increasing in our surroundings due to enhanced pollution levels. It has become essential to put the enhanced level of heavy metal contamination under vigilance as it has not yet risen to a dangerous level at the moment. The consumption of food and potable water generally affects the levels of these elements in human specimens. But, there is a danger of buildup of small doses either through inhalation, bioaccumulation or skin absorption that can lead to unpleasant consequences. In order to create a healthy and equitable living environment for future generations, we should rather reduce or minimize the circle of poison of these heavy metals (As, Cd, Hg, and Pb) and take immediate steps to limit human exposure to heavy metals.

KEYWORDS

- heavy metals
- AAS
- heavy metal toxicity
- quantitative estimation
- statistical analysis

REFERENCES

1. Rathi, S. Alternative Approaches for Better Municipal Solid Waste Management in Mumbai, India. *Waste Manage.* **2006**, *26*, 1192–1200.
2. Boyd, R. S. Heavy Metal Pollutants and Chemical Ecology: Exploring New Frontiers. *J. Chem. Ecol.* **2010**, *36*, 46–58.
3. Valko, M.; Rhodes, C. J.; Moncol, J.; Izakovic, M. Free Radicals, Metals and Antioxidants in Oxidative Stress-induced Cancer. *Chem.–Biol. Interact.* **2006**, *160*, 1–40.
4. Issazadeh, K.; Jahanpour, N.; Pourghorbanali, F.; Raeisi, G.; Faekhondeh, J. Heavy Metals Resistance by Bacterial Strain. *Ann. Biol. Res.* **2013**, *4*, 60–63.
5. Morais, S.; Costa Garcia, F. G.; Pereira, M. L. Heavy Metals and Human Health. *Environ. Health—Emerg. Iss. Prac.* **2012**, *10*, 228–246.
6. Suruchi; Khanna, P. Assessment of Heavy Metal Contamination in Different Vegetables Grown in and Around Urban Areas. *Res. J. Environ. Toxicol.* **2011**, *5*, 162–179.
7. Agency for Toxic Substances and Disease Registry, 4770 Buford Hwy NE Atlanta, GA 30341. (Oct 2011 update).
8. Status of Trace and Toxic Metals in Indian Rivers, Government of India, Ministry of Water Resources, Central Water Commission, May 2014.
9. Chaurasia, N.; Mishra, A.; Pandey, S. K. Finger Print of Arsenic Contaminated Water in India—A Review. *J. Foren. Res.* **2012**, *3* (10), 1–4.
10. Dey, T. K.; Bakhshi, M.; Banerjee, P.; Ghosh, S. Groundwater Arsenic Contamination in West Bengal: Current Scenario, Effects and Probable Ways of Mitigation. *Int. Lett. Nat. Sci.* **2014**, *13*, 45–58.
11. Bhujal News, Arsenic in Groundwater in India, April–Sept 24, 2009.

12. Lead Toxicity, ENVIS-NIOH News Letter, Oct–Dec 4, 2009.
13. Farid, S. M.; Enani, M. A.; Wajid, S. A. Determination of Trace Elements in Cow Milk in Saudi Arabia. *J. King Saud Univ. Eng. Sci.* **2004,** *15,* 131–140.
14. Bush, V. J.; Moyer, T. P.; Batts, K. P.; Parisi, J. E. Essential and Toxic Element Concentrations in Fresh and Formalin-Fixed Human Autopsy Tissues. *Clin. Chem.* **1995,** *41* (2), 284–294.
15. Exley, K. Human Biomonitoring to Assess Environmental Chemical Exposure: Work Towards UK Framework. *Persp. Public Health* **2014,** *134* (5), 299.
16. ToxGuide for Arsenic, U.S. Department of Health and Human Services, Public Health Service, Agency for Toxic Substances and Disease Registry.
17. ToxGuide for Cadmium, U.S. Department of Health and Human Services, Public Health Service, Agency for Toxic Substances and Disease Registry.
18. ToxGuide for Lead, U.S. Department of Health and Human Services, Public Health Service, Agency for Toxic Substances and Disease Registry.
19. ToxGuide for Mercury, U.S. Department of Health and Human Services, Public Health Service, Agency for Toxic Substances and Disease Registry.
20. Vaughan David, J. Arsenic. *Elements* **2006,** *2,* 71–75.
21. Basu, A.; Sen, P.; Jha, A. Environmental Arsenic Toxicity in West Bengal, India: A Brief Policy Review. *Ind. J. Public Health* **2015,** *59* (4), 295–298.
22. Adhikary, R.; Mandal, V. Status of Arsenic Toxicity in Ground Water in West Bengal, India: A Review. *MOJ Toxicol.* **2017,** *3* (5), 1–5.
23. Bullar, D. S.; Thind, A. S.; Singla, A. Childhood Lead Poisoning—A Review. *J. Punjab Acad. Forensic Med. Toxicol.* **2015,** *15* (1), 43–49.
24. World Health Organization. *Childhood Lead Poisoning*; World Health Organization, 2010.
25. World Health Organization. *International Lead Poisoning Prevention Week of Action*; World Health Organization, Oct 2013.
26. Bhowmik, D.; Kumar, K.; Sampath, P.; Umadevi, M. Lead Poisoning- the Future of Lead's Impact Is Alarming on Our Society. *The Pharma Innov.* **2012,** *1,* 40–49.
27. Sharma, R.; Sharma, R. K.; Chaudhary, A.; Lalita. Assessment of Physico—Chemical Properties of Ground Water Samples Collected from Jaipur, Rajasthan, India. *Int. Res. J. Environ. Sci.* **2017,** *6* (7), 16–22.
28. Godt, J.; Scheidig, F.; Grosse-Siestrup, C. The Toxicity of Cadmium and Resulting Hazards for Human Health. *J. Occup. Med. Toxicol.* **2006,** *1,* 1–22.
29. Chunhabundit, R. Cadmium Exposure and Potential Health Risk from Foods in Contaminated Area, Thailand. *Toxicol. Res.* **2016,** *32,* 65–72.
30. Counter, S. A.; Buchanan Leo, H. Mercury Exposure in Children: A Review. *Toxicol. Appl. Pharmacol.* **2004,** *198* (2), 209–230.
31. Pirrone, N.; Cinnirella, S.; Feng, X.; Finkelman, R. B.; Friedli, H. R.; Leaner, J.; Mason, R.; Mukherjee, A. B.; Stracher, G. B.; Streets, D. G.; Telmer, K. Global Mercury Emissions to the Atmosphere from Anthropogenic and Natural Sources. *Atmos. Chem. Phys.* **2010,** *10,* 5951–5964.

CHAPTER 6

Shedding Light on New Drug Designs by Modeling Peptide Structures via Quantum Chemical Methods in the Health Field

SEFA CELIK[1*], SERDA KECEL-GUNDUZ[1], SEVIM AKYUZ[2], and AYSEN E. OZEL[1]

[1]Physics Department, Science Faculty, Istanbul University, Vezneciler, Istanbul 34134, Turkey

[2]Physics Department, Science and Letters Faculty, Istanbul Kultur University, Atakoy Campus, Bakirkoy, Istanbul 34156, Turkey

*Corresponding author. E-mail: scelik@istanbul.edu.tr

ABSTRACT

Peptides and proteins have great potential as therapeutics. Peptides are considered highly selective and efficacious and, at the same time, are recognized for being relatively safe and well tolerated. For this reason, peptides promise to be the starting point for drug development. The biological function and activity of peptides and proteins are correlated using conformational properties. Peptide activity is related to both the presence of functionally active groups that are bound to a target protein and the conformational properties of the whole peptide. Conformational flexibility, which is investigated by computational methods, gives substantial opportunities for new rational drug design. The behavior of molecular systems is simulated by various computational methods through conformational analysis, molecular dynamics (MDs), potential energy surface (PES), etc., and performed by theoretical molecular modeling methods. The conformational analysis method is a preferred technique for analyzing the peptide structure, which is carried out by conformational energy calculations concerning the spatial location of the peptide side and backbone chains. MD techniques determine not only the conformational variation of peptide molecules, which are used as an ingredient of a drug, but also the information where drug molecules bind together, and how they exert their effects. MD is also an applicable tool in understanding solvent effects on peptide conformation. Because of the information derived from simulations of MDs, it can ensure new insights at the molecular level for different biological systems. The present and future of structure-based drug discovery will make use of the advantages of MD.

PES is a method that determines the relationship between molecular structure and energy and has a key role in molecular structure studies. Molecular docking, which comprises different types of algorithms such as MDs and Monte Carlo simulations, is another technique used to explain the preferred binding modes of ligand–receptor complexes, where the ligand is usually a small molecule and the receptor is a protein. Since the interaction between the ligand and the receptor allows us to predict the activation or inhibition of protein, such information may be used for drug design.

6.1 INTRODUCTION

Determination of the biochemical and physicochemical properties of molecules by using a computer is called molecular simulation. Computer simulations that function as a bridge between theoretical and experimental results are very important. Information about macroscopic properties of the system can be obtained with computer simulations by starting from its microscopic details. The simulations can be carried out perfectly under conditions of very high temperature or pressure, whereas performing experiments is very difficult. Furthermore, the rates of molecular phenomena, which are difficult to examine in terms of experimental investigation, are possible to examine with simulations. The basis of the method dates back to the late 1950s,[1–3] and physical, chemical, and biological phenomena can be studied using computer models over a wide spectrum from molecular scales to galaxies.[4]

Using a simulation technique:
- The structures and geometries of molecules can be investigated and displayed.

- The energies of the molecules can be calculated.
- Different conformations of the molecule and the energies of these conformations can be determined.
- Quantities, such as boiling point, molar volume, density, thermodynamics, magnetic susceptibility, dipole moment, electrostatic potential, and accessible surface area, can be calculated by interpolation, extrapolation, or direct calculation methods.

Amino acids, the smallest building blocks of peptides and proteins have significant roles in metabolism.[5-9] The structure of amino acids contains an amino group, a carboxyl group, a Cα carbon atom, an H atom, and a variable group. The name of the amino acid changes according to the variable group. The general structure of amino acids is given in Figure 6.1.

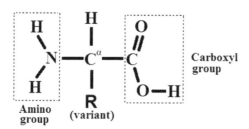

FIGURE 6.1 The general structure of amino acids.

Peptides are molecules formed by bringing up to 20 amino acids together—which are also called standard amino acids—in different sequences and with different numbers of peptide bonds. The general structure of a peptide chain is shown in Figure 6.2.

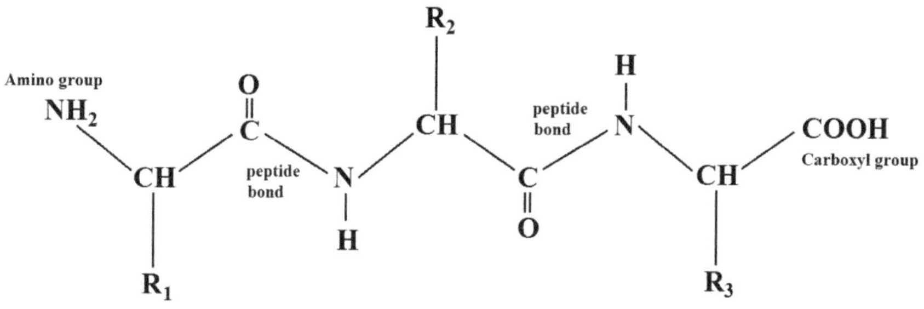

FIGURE 6.2 The peptide structure and its bonds.

These peptide structures have many significant biological activities, such as antimicrobial,[10–13] anticancer,[14–16] antihypertension,[17–20] anti-inflammatory,[21–24] analgesic,[25–28] neuroprotective,[29–32] antioxidant,[33–36] and antiangiogenic.[37–40] Therefore, there are numerous studies in literature on the determination of the structures of peptides.[41–61] Because there is a relationship between the structures and activities of these biologically important molecules, the determination of the most stable geometries of these molecules is of great significance for drug designs.[62–69] Thus, the studies concentrate on investigating the most stable structure using various calculation methods.

Conformational analysis is based on the determination of the lowest energy of the molecular system by repeating the rotation of intramolecular bonds and angles (dihedral angles).

Methods, such as conformational analysis, molecular dynamics (MDs), and potential energy surface (PES) determinations, are some of the methods used to find the most stable structure. Although the purpose of all these methods is to find the most stable geometry, their working principles differ from each other.

6.2 MATERIALS AND METHOD

6.2.1 CONFORMATIONAL ANALYSIS METHOD

6.2.1.1 THEORETICAL CONFORMATIONAL ANALYSIS METHOD

Molecular conformation involves temporary molecular shapes that occur in a molecule produced by the rotation of constituent groups around the sigma bond (single bond). The analysis of the energy change formed due to the rotation of the groups around the sigma bond is called conformational analysis. Different conformations of molecules may result in different properties in the molecules, which are usually low-energy conformations. Conformational analysis is needed to find the conformation of large biomolecules with low energies. Conformational analysis diagram is given in Figure 6.3.

The most stable amino acid structures are determined by Ramachandran maps. The most stable peptide structures are obtained with conformational analysis by rotating the dihedral angles on the main and side chains.[70,71]

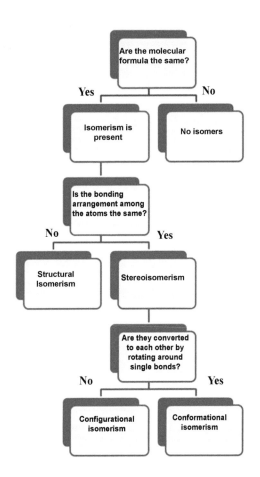

FIGURE 6.3 Conformational analysis diagram.

The conformation analysis program created by Maksumov et al. also makes calculations based on these approaches.[72] Based on Ramachandran maps, each conformation energy of the peptide is obtained by calculations using hydrogen bond, van der Waals, electrostatic and torsion energy contributions, and the dielectric constant of the medium. The van der Waals interaction is calculated using the Lennard-Jones potential with the parameters proposed by Scott and Scheraga.[73,74] Electrostatic energy is calculated according to Coulomb's law defined by the charge values and distances suggested in Reference [75]. The accepted values for the torsion potentials and barrier heights are defined in Reference [76].

A study of the Arg-Lys-Asp-Val tetrapeptide used the conformational analysis program[72] to determine the 10 lowest energy conformers and the most stable structure among them (see Figs. 6.4 and 6.5); the contributions of each conformation to the total energy are shown in Table 6.1.[77]

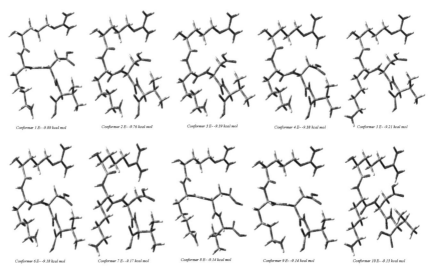

FIGURE 6.4   Ten most stable conformations obtained from the conformational analysis of the Arg-Lys-Asp-Val tetrapeptide.[77]

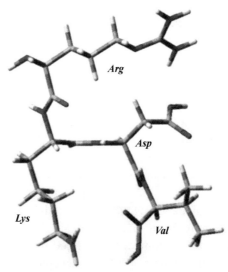

FIGURE 6.5   The most stable structure of the Arg-Lys-Asp-Val tetrapeptide (Conformer 1).[77]

TABLE 6.1 The Conformational Potential Energy of the Global Conformation of the Arg-Lys-Asp-Val Tetrapeptide.[77]

	Total energy (kcal/mol)	van der Waals (kcal/mol)	Eelectrostatic (kcal/mol)	Etorsion (kcal/mol)
1	-9.89	-16.75	4.64	2.23
2	-9.76	-16.21	4.31	2.14
3	-9.39	-15.90	4.28	2.24
4	-9.38	-16.93	4.72	2.83
5	-9.21	-16.58	4.47	2.90
6	-9.18	-16.84	4.64	3.02
7	-9.17	-16.95	5.25	2.52
8	-9.14	-16.46	4.29	3.02
9	-9.14	-16.42	4.52	2.75
10	-9.13	-17.35	4.94	3.28

In a study conducted in 2010, 2011, and 2015, the most stable geometries of the tachykinin, neuropeptide, eledoisin, the antihypertensive peptide ovokinin, and the human and mouse hemokinin-1 molecules were obtained using theoretical conformational analysis.[78–80]

Two common techniques are used in computer simulations of atom and molecules: the Monte Carlo (MC) and the MD techniques.

- The MC method is an implementation of the Metropolis method.[81]
- The MD method is based on the integration of Newton's second law of motion.[82]

6.2.1.2 MOLECULAR DYNAMIC METHOD

MD is a numerical method for studying many–particle systems such as molecules, clusters, and even macroscopic systems such as gases, liquids, and solids. There are two different classes of MD simulations; classical MD and ab initio MD. In the case of classical MDs, the interactions are approximated using the classical empirical potentials and lead to simulation of purely classical many–particle problems. On the other hand, the ab initio MDs perform a full quantum calculation of the electronic structure, at every time step (for every configuration of the atomic nuclei). The forces are found from the dependence of the energy on the particle positions. It has not only much higher accuracy than classical MD, but also much higher numerical effort (restricts a number of particles and simulation time).

The energy function allows us to calculate the force experienced by any atom given the positions of the other atoms. Newton's laws tell us how those forces will affect the motions of the atoms.

Input information is necessary to make an MD calculation. This input information includes the potential energy function and the first position of each atom. The potential energy function is a mathematical expression that varies according to atomic structure and type. The gradient of this expression gives the value of the force on each atom. By using the force value on each atom, the accelerations of each atom in the system are determined. With the help of the equation of motion, the acceleration, velocity, and orbit of the particle over time are obtained for each atom. Atoms are then placed according to the velocity information obtained for their new positions, and this phenomenon is continued until the total potential energy for the cycle calculation system becomes constant by considering the potential energy function. Through this method, the state of the system can be determined at any time in the past or the future, if the velocity and position of each atom is known at any moment. Figure 6.6 shows the working mechanism of the MD system.

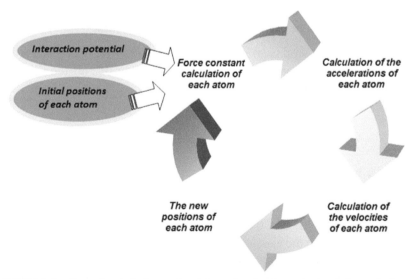

FIGURE 6.6   General calculation cycle of the molecular dynamic simulation technique.

The force fields used for molecular systems can be categorized as having two basic components: internal and external forces. There are some restrictions on energy calculations that are related to the deviations in their angular values—formed between atoms—from the equilibrium and reference values. In the force field function, these angular expressions describe the energy change of the system when bonds rotate and bend. In the more complex force field function, there are expressions for bond stretching, angle bending, dihedral rotations, and electrostatic interactions between the systems that do not bond with each other.

6.2.1.2.1 Bond Stretching Potential Energy

The interaction between two atoms is determined by the bond length-dependent potential, and this potential is usually a second-order equation. It can be considered to be like a spring physically placed between two atoms. The coefficient of the function, which depends on the bond type and the atoms it contains, is equal to the spring constant.

$$E_{\text{bond}} = \sum_{i,j} k_{i,j}{}^{b} \left(r_{i,j} - r_{i,j}{}^{0} \right)^{2}$$

where $k_{i,j}{}^{b}$ force constant and $r_{i,j}$: the deviation of the bond from equilibrium.

6.2.1.2.2 Angle Bending Potential Energy

When atoms are linked by chemical bonds, they also form an angle between the two bonds. This angular interaction is expressed as the angular bond potential and written as a second-order equation. This potential is associated with changes in the bond angles.

$$E_{\text{angle}} = \sum_{i,j} k_{i,j,k}{}^{a} \left(\theta_{i,j,k} - \theta_{i,j,k}{}^{0} \right)^{2}$$

where $k_{i,j,k}{}^{a}$: force constant, $\theta_{i,j,k}{}^{0}$: the equilibrium angle, and $\theta_{i,j,k}$: the angle at any time.

6.2.1.2.3 Torsional Potential Energy

To determine the spatial structure of a molecule in the given valence bond, and the valence angle values, it is also necessary to know the dihedral angle values that indicate the spatial position of the main and side chains of the molecule. The molecule rotates around single chemical bonds with a small amount of energy taken from the media. This motion is not completely free; there is a barrier to rotation. The torsional energy is defined as the Lennard-Jones potential.

$$E_{\text{tor}} = \sum_{i,j,k,l} k_{i,j,k,l}{}^{c} \left[1 \pm \cos n \left(\theta_{i,j,k,l} - \theta_{i,j,k,l}{}^{0} \right) \right]$$

where $k_{i,j,k,l}{}^{c}$: force constant between atoms, $\theta_{i,j,k,l}{}^{0}$: the equilibrium angle, and $\theta_{i,j,k,l}$: the angle at any time.

6.2.1.2.4   Electrostatic Interaction Potential Energy

This potential is used to calculate the potential energy between two charged atoms. While this potential is inversely proportional to the distance between the two atoms, it is directly proportional to the charges on the atoms.

$$E_{\text{elec}} = \sum_{i,j}^{N} \frac{q_i q_j}{\varepsilon r_{ij}}$$

where q_i and q_j: the charges on the atoms, r_{ij}: the distance between atoms, and e: the dielectric constant of the media.

6.2.1.2.5   van der Waals Interaction Potential Energy (Nonbonded Interaction Energy)

Although the arrangement of electrons around an atom is symmetrical, a constant fluctuation in the electron distribution implies that the arrangement can be asymmetrical for a while. Such a situation causes atoms to attract each other because it forms an instant dipole and induces an interaction in all neighboring atoms. This is valid for all atoms, even for noble gases. However, when two atoms approach each other very closely, the repulsion between electron clouds is greater than the induced attraction. There is an optimum distance between two nonbonded atoms known as the van der Waals contact distance. The repulsion and attraction forces at this distance are known as the London dispersion forces and these weak bonds resulting from dipole–dipole, dipole-induced dipole interactions are called van der Waals bonds. They play an important role in maintaining the continuity of the 3D structure created by the protein. An ideal van der Waals interaction occurs when the distance between atoms is between 2.5 and 2.7 Å. Although the van der Waals interaction is less than 1/100 of the chemical bond in terms of energy, its contribution to the energy of the molecule becomes significant because it occurs among many atoms. The Lennard-Jones potential was used to describe this energy.

$$E_{vw} = \sum_{i,j} \varepsilon_{i,j} \left[\left(\frac{\sigma_{ij}}{r_{ij}} \right)^{12} - 2 \left(\frac{\sigma_{ij}}{r_{ij}} \right)^6 \right]$$

In MD calculations, the total energy of the system is calculated as follows:

$$E_{total} = E_{bond} + E_{angle} + E_{torsion} + E_{electrostatic} + E_{vwv}$$

The most stable geometries for some peptide structures were obtained using MD simulation by Mehrazma et al. in 2017.[83]

Stable conformations of $A\beta_{1-16}$, $A\beta_{1-16}$, isoD7-$A\beta_{1-16}$, and $A\beta_{1-16}$ peptides were determined by Andrey et al. in 2016.[84]

Phe-Tyr dipeptide that was investigated in Wakame food with the greatest ACE-inhibitory activity is used as a pharmaceutical drug for the treatment of hypertension, cardiovascular diseases, and diabetic nephropathy.[85] In the literature, there are no complete studies that have investigated the effects of Phe-Tyr dipeptide, in the treatment of hypertension and cardiovascular diseases. Conformational design, characterization, and in vitro studies of this dipeptide are needed to determine the effects of ACE-inhibitory activity on blood pressure. Because of that, the molecular structure of Phe-Tyr dipeptide was identified by using MD calculations, including those related to determining the dipeptide's soluble effect. MD calculations are preferred for investigating the conformational change of the peptide molecule.[86] This simulation technique is also a valid tool for understanding protein folding and unfolding dynamics and demonstrates any significant interactions between protein atoms and solvent.[87] The starting dipeptide structure was generated placing an ideal geometry into a triclinic box of SPC (simple point charge) water molecules with 2906 water, and counterions (Na+ and Cl-) are added to neutralize the system.

The docking calculations made using the AutoDock Vina program gave nine conformers at the lowest energy that can bind to the active region of the protein, and their binding energies. It was seen that the conformer ranked first (-8.5 kcal/mol) had the most stable geometry.

6.2.1.3   POTENTIAL ENERGY SURFACE METHOD

The PES, an important concept in theoretical chemistry, is a significant mathematical function that gives the energy of the molecule as a function of conformation. In particular, it is based on the systematic rotation of flexible dihedral angles at regular intervals. During this process, the rotation continues until the molecule reaches its first geometry and the energy of each conformer is calculated. From this, the most stable geometry is then determined. The minimum and saddle points can be defined by plotting the graph indicating the change in dihedral angle and energy. In a study performed on cycloheptapeptide in 2017, 1024 conformations were investigated using the AM1 method by rotating five dihedral angles simultaneously.[46]

The PES of the *N*-acetyl *O*-methoxy (*N*-acetyl OMe) proline molecule was calculated based on distance by Istvan et al. in 2005 using the DFT method; and local and global minimum values were obtained.[88]

In a study conducted on PES in 2015, the metatyrosine molecule was investigated with the M06-2X/6-311++G(d,p) level of theory.[89]

6.2.1.4   DOCKING METHOD

The molecular docking problem is finding the solution to a "key–lock problem," where a ligand is the key and a protein is the lock.[90] The docking method is mostly used to determine the interaction and activity of small molecules that are candidates for medicine used with protein targets. Today, there are more sensitive docking methods where both protein and ligand undergo flexible and conformational changes. This can be associated with the glove–hand analogy rather than the key–lock analogy.[91] For the docking process, docking studies—where conformational adjustments of both the protein surface and ligand are made—give more accurate results.

The molecular docking method is used to make calculations with macromolecules, proteins, and enzymes. It is a molecular modeling method in which the most probable binding position that a molecule will prefer is determined when forming a stable complex with another molecule.[92] The protein is thought of as the lock and the ligand is thought of as the key in terms of bonding. In the docking process, depending on the interaction between the "ligand" and the active site of the receptor, the binding energies and conformation of the ligand can be determined with respect to binding energy, and the most stable conformation of the ligand–receptor interaction is obtained.[93]

The MD simulations are used to determine where the peptide molecule binds to its receptor, and also to investigate how the binding strength of peptide changes if it binds elsewhere by changing the protein's structure.

The simulations in which a receptor protein transitions spontaneously from its active structure to its inactive structure are used to describe the mechanism by which drugs binding to one end of the receptor cause the other end of the receptor to change shape (activate).

There are two basic docking approaches. The first one is the morphological coherency based on the complementary shape and topology by defining the surface of the protein.[94,95] The second one is the simulation method based on the determination of the binary binding energies in the specified conformational spaces.[96]

- *Constant receptor—constant ligand docking:* This is very rapid but has low accuracy; it is effective in scanning very large ligand libraries.
- *Constant receptor—flexible ligand docking:* This is obtained by considering the rotating ligand bonds in the docking calculation.
- *Flexible receptor—flexible ligand docking:* This is obtained by considering the rotation angles of the amino acids within the protein binding site of the side chains, in the binding calculations. It requires more calculation power but has higher accuracy.

Most docking programs in use account for the whole conformational space of the ligand (flexible ligand), and several attempt to model a flexible protein receptor. Each "snapshot" of the ligand–protein pair is referred to as a "pose." Different docking programs may give different results for ligand arrangements and binding poses calculations because they use different algorithms. Therefore, it is advantageous to make calculations with more than one docking program and then analyze them by looking at common results. While the binding energies produced by docking programs are not exact, binding poses can often be accurately determined. The most important result in docking calculations is the ordering of ligands: the higher order ligands are more likely to bind, while the lower order ligands have lower binding potential.

As a result of the docking calculation studies done on cycloheptapeptide–fibronectin in 2017, the most stable conformation of the peptide and the binding sites in the active region of the fibronectin were determined.[46]

The docking calculation for cathepsin B from *Hordeum vulgare* and Aβ peptide was done by Maruti et al. in 2016 and the complex structure was defined.[97]

In a study conducted in 2017,[85] the interaction of the peptide with the protein systems was also determined by calculating the free energy of binding using the molecular docking technique, a method used to define a ligand that binds to a specific receptor binding site and to ascertain its preferred, energetically most favorable binding pose.[98] To improve the bioavailability of Phe-Tyr, a delivery system based on poly(lactic-*co*-glycolic acid) (PLGA) nanoparticles (NPs) loaded with Phe-Tyr (Phe-Tyr-PLGA NPs) for treating hypertension and cardiovascular diseases was prepared in this study. Encapsulation with PLGA may not only useful for delivery of Phe-Tyr, but also may increase biocompatibility. This study presents properties of Phe-Tyr-PLGA NPs drug molecule that can provide insights for improved new drug design and formulation for the treatment of hypertension disease. Hence, the entire studies will help in improvising the drug designing with better reactivity and stability. The schematic form for the docked conformation of active site of Phe-Tyr-PLGA NPs drug molecule (1), schematic hydrogen bonding for the docked conformation of active site of Phe-Tyr-PLGA NPs drug molecule (2), schematic hydrogen bonding for the docked conformation of active site of Phe-Tyr-PLGA NPs drug molecule with TYR35 (3), and ARG167 (4) were shown.

The significance of protein–ligand docking methods in drug design is indisputable. In drug design, the docking studies performed on the interactions between small molecules and receptors have become increasingly important. Some applications that have been designed by simulations and used in a variety of treatments are presented in Table 6.2.

6.3 CONCLUSION

Nowadays, in silico techniques (performed on computer or via computer simulation) are much preferred to those that produce biologically active compounds at a high cost using organic synthesis methods. The goal of molecular modeling is to understand the relationship between the chemical and physical properties of a molecule, its chemical structure, and its three-dimensional geometry. By correlating these properties using calculations based on the chemical and biological functions of the molecules, it is possible to predict the most suitable designs. As a result of modeling the

TABLE 6.2 Some of the Simulation Applications Used in Drug Design.

Drug	Therapeutic area	Pharmaceutical company	Method	References
TEVETEN® Eprosartan	Antihypertensive	Smithkline Beecham	Molecular modeling	[99,100]
ZOMIG® Zolmitriptan	Migraine	Wellcome, Zeneca	Molecular modeling	[101]
CRIXIVAN® Indinavir	AIDS	Merck Company	Molecular mechanics for calculation	[102,103]
TRUSOPT® Dorzolamide	Glaucoma	Merck Company	Ab initio calculations	[104,105]
NOVLADEX® Tamoxifen	Anticancer	Generics (UK) Ltd	Quantum and molecular mechanics	[106]
GEMZAR® Gemcitabines	Anticancer	Hospira UK Ltd	Molecular dynamics	[107]
PERJETA™ Pertuzumab	Anticancer	Genentech	Molecular dynamics	[108]
ZINECARD® Dexrazoxane	Chemoprotectantive agent	Pfizer Ltd	Molecular mechanics	[109]
VIAGRA® Sildenafil	Treatment of erectile dysfunction and pulmonary arterial hypertension	Pfizer Ltd	Molecular modeling	[110,111]

peptide structures using appropriately and correctly selected quantum chemical methods, increasing the activity of known drugs, designing new drug molecules, the structure of which is unknown, or further development of existing drugs are becoming more important day by day. Saving time by obtaining data faster and easier is one of the most important advantages of modeling. The relationship between chemical structure and function is of significant use in many areas of industry, including molecular modeling, molecular biology, protein science, drug design, electronics and photonic materials, and polymer science.

KEYWORDS

- peptide
- conformational analysis
- quantum chemical methods
- molecular dynamics
- molecular modeling

REFERENCES

1. Alder, B. J.; Wainwright, T. E. Studies in Molecular Dynamics. I. General Method. *J. Chem. Phys.* **1959,** *31* (2), 459–466.
2. Alder, B. J.; Wainwright, T. E. Studies in Molecular Dynamics. II. Behavior of a Small Number of Elastic Spheres. *J. Chem. Phys.* **1960,** *33* (5), 1439–1451.
3. Wainweight, T.; Alder, B. J. Molecular Dynamics Computations for the Hard Sphere System. *Il Nuovo Cimento (1955–1965)* **1958,** *9,* 116–132.
4. Allen, M. P.; Tildesley, D. J. *Computer Simulation of Liquids*; Clarendon Press: Oxford, 1997.
5. Oh, J. G.; Chun, S. H.; Kim, J. H.; Shin, H. S.; Cho, Y. S.; Kim, Y. K.; … Lee, K. W. Anti-Inflammatory Effect of Sugar-Amino Acid Maillard Reaction Products on Intestinal Inflammation Model In Vitro and In Vivo. *Carbohydr. Res.* **2017,** *449,* 47–58.
6. Tsun, Z. Y.; Possemato, R. Amino Acid Management in Cancer. In *Seminars in Cell & Developmental Biology* (Vol. 43); Academic Press, 2015; pp 22–32.
7. Wu, G. Functional Amino Acids in Growth, Reproduction, and Health. *Adv. Nutr. Int. Rev. J.* **2010,** *1* (1), 31–37.

8. Van De Poll, M. C.; Soeters, P. B.; Deutz, N. E.; Fearon, K. C.; Dejong, C. H. Renal Metabolism of Amino Acids: İts Role in İnterorgan Amino Acid Exchange. *Am. J. Clin. Nutr.* **2004**, *79* (2), 185–197.

9. Volpi, E.; Kobayashi, H.; Sheffield-Moore, M.; Mittendorfer, B.; Wolfe, R. R. Essential Amino Acids Are Primarily Responsible for the Amino Acid Stimulation of Muscle Protein Anabolism in Healthy Elderly Adults. *Am. J. Clin. Nutr.* **2003**, *78* (2), 250–258.

10. Wang, G. Human Antimicrobial Peptides and Proteins. *Pharmaceuticals* **2014**, *7* (5), 545–594.

11. Schauber, J.; Gallo, R. L. Antimicrobial Peptides and the Skin İmmune Defense System. *J. Allergy Clin. Immunol.* **2008**, *122* (2), 261–266.

12. Brogden, K. A. Antimicrobial Peptides: Pore Formers or Metabolic İnhibitors in Bacteria? *Nat. Rev. Microbiol.* **2005**, *3* (3), 238–250.

13. Valore, E. V.; Park, C. H.; Quayle, A. J.; Wiles, K. R.; McCray Jr, P. B.; Ganz, T. Human Beta-Defensin-1: An Antimicrobial Peptide of Urogenital Tissues. *J. Clin. Invest.* **1998**, *101* (8), 1633.

14. Gaspar, D.; Veiga, A. S.; Castanho, M. A. From Antimicrobial to Anticancer Peptides. A Review. *Front. Microbiol.* **2013**, *4*, 1–16.

15. Davitt, K.; Babcock, B. D.; Fenelus, M.; Poon, C. K.; Sarkar, A.; Trivigno, V.; … Shaikh, M. F. The Anti-Cancer Peptide, PNC-27, Induces Tumor Cell Necrosis of a Poorly Differentiated Non-Solid Tissue Human Leukemia Cell Line that Depends on Expression of HDM-2 in the Plasma Membrane of These Cells. *Ann. Clin. Lab. Sci.* **2014**, *44* (3), 241–248.

16. Crunkhorn, S. Anticancer Drugs: Stapled Peptide Rescues p53. *Nat. Rev. Drug Discov.* **2011**, *10* (1), 21.

17. Norris, R.; FitzGerald, R. J. Antihypertensive Peptides from Food Proteins. In *Bioactive Food Peptides in Health and Disease*; InTech, 2013.

18. Martínez-Maqueda, D.; Miralles, B.; Recio, I.; Hernández-Ledesma, B. Antihypertensive Peptides from Food Proteins: A Review. *Food Func.* **2012**, *3* (4), 350–361.

19. Jakala, P.; Vapaatalo, H. Antihypertensive Peptides from Milk Proteins. *Pharmaceuticals* **2010**, *3* (1), 251–272.

20. Hong, F.; Ming, L.; Yi, S.; Zhanxia, L.; Yongquan, W.; Chi, L. The Antihypertensive Effect of Peptides: A Novel Alternative to Drugs? *Peptides* **2008**, *29* (6), 1062–1071.

21. Wei, L.; Huang, C.; Yang, H.; Li, M.; Yang, J.; Qiao, X.; … Xu, W. A Potent Anti-Inflammatory Peptide from the Salivary Glands of Horsefly. *Parasites Vectors* **2015**, *8* (1), 556.

22. Wei, E. T.; Thomas, H. A. Anti-Inflammatory Peptide Agonists. *Ann. Rev. Pharmacol. Toxicol.* **1993**, *33* (1), 91–108.

23. Miele, L.; Cordella-Miele, E.; Facchiano, A.; Mukherjee, A. B. Novel Anti-Inflammatory Peptides from the Region of Highest Similarity Between Uteroglobin and Lipocortin I. *Nature* **1988**, *335* (6192), 726–730.

24. Briggs, J. B.; Larsen, R. A.; Harris, R. B.; Sekar, K. V.; Macher, B. A. Structure/Activity Studies of Anti-İnflammatory Peptides Based on a Conserved Peptide Region of the Lectin Domain of E-, L- and P-Selectin. *Glycobiology* **1996**, *6* (8), 831–836.

25. Aldrich, J. V.; McLaughlin, J. P. Opioid Peptides: Potential for Drug Development. *Drug Discov. Today Technol.* **2012**, *9* (1), e23–e31.

26. Stein, C.; Hassan, A. H.; Lehrberger, K.; Giefing, J.; Yassouridis, A. Local Analgesic Effect of Endogenous Opioid Peptides. *Lancet* **1993**, *342* (8867), 321–324.

27. Lee, H. K.; Zhang, L.; Smith, M. D.; Walewska, A.; Vellore, N. A.; Baron, R.; … Bulaj, G. A Marine Analgesic Peptide, Contulakin-G, and Neurotensin Are Distinct Agonists for Neurotensin Receptors: Uncovering Structural Determinants of Desensitization Properties. *Front. Pharmacol.* **2015**, *6,* 1–12.

28. De Lima, M. E.; Borges, M. H.; Verano-Braga, T.; Torres, F. S.; Montandon, G. G.; Cardoso, F. L.; … Benelli, M. T. Some Arachnidan Peptides with Potential Medical Application. *J. Venom. Anim. Toxins İncl. Trop. Dis.* **2010**, *16* (1), 8–33.

29. Caruso, C.; Carniglia, L.; Durand, D.; Scimonelli, T. N.; Lasaga, M. Melanocortins: Anti-Inflammatory and Neuroprotective Peptides. In *Neurodegeneration*; InTech, 2012.

30. Gozes, I. Neuroprotective Peptide Drug Delivery and Development: Potential New Therapeutics. *Trends Neurosci.* **2001**, *24* (12), 700–705.

31. Patocka, J.; Slaninová, J.; Kunešová, G. Neuroprotective Peptides as Drug Candidates Against Alzheimer's Disease. *J. Appl. Biomed.* **2005**, *3*, 67–73.

32. Sari, Y.; Segu, Z. M.; YoussefAgha, A.; Karty, J. A.; Isailovic, D. Neuroprotective Peptide ADNF-9 in Fetal Brain of C57BL/6 Mice Exposed Prenatally to Alcohol. *J. Biomed. Sci.* **2011**, *18* (1), 77.

33. Elias, R. J.; Kellerby, S. S.; Decker, E. A. Antioxidant Activity of Proteins and Peptides. *Crit. Rev. Food Sci. Nutr.* **2008**, *48* (5), 430–441.

34. Wang, X. J.; Zheng, X. Q.; Kopparapu, N. K.; Cong, W. S.; Deng, Y. P.; Sun, X. J.; Liu, X. L. Purification and Evaluation of a Novel Antioxidant Peptide from Corn Protein Hydrolysate. *Proc. Biochem.* **2014**, *49* (9), 1562–1569.

35. Fan, J.; He, J.; Zhuang, Y.; Sun, L. Purification and Identification of Antioxidant Peptides from Enzymatic Hydrolysates of Tilapia (*Oreochromis niloticus*) Frame Protein. *Molecules* **2012**, *17* (11), 12836–12850.

36. Chen, H. M.; Muramoto, K.; Yamauchi, F.; Nokihara, K. Antioxidant Activity of Designed Peptides Based on the Antioxidative Peptide İsolated from Digests of a Soybean Protein. *J. Agric. Food Chem.* **1996**, *44* (9), 2619–2623.

37. Rosca, E. V.; Koskimaki, J. E.; Rivera, C. G.; Pandey, N. B.; Tamiz, A. P.; Popel, A. S. Anti-Angiogenic Peptides for Cancer Therapeutics. *Curr. Pharm. Biotechnol.* **2011**, *12* (8), 1101–1116.

38. Rüegg, C.; Hasmim, M.; Lejeune, F. J.; Alghisi, G. C. Antiangiogenic Peptides and Proteins: From Experimental Tools to Clinical Drugs. *Biochim. Biophys. Acta* **2006**, *1765* (2), 155–177.

39. Chan, L. Y.; Craik, D. J.; Daly, N. L. Dual-Targeting Anti-Angiogenic Cyclic Peptides as Potential Drug Leads for Cancer Therapy. *Sci. Rep.* **2016**, *6*, 35347.

40. Yamada, K.; Malik, A. B. Anti-Angiogenic Peptide Derived from Kinesin KIF13B Inhibits Trafficking of VEGFR2 to Endothelial Cell Surface. *FASEB J.* **2017**, *31* (1 Supplement), 990–998.

41. Kecel, S.; Ozel, A. E.; Akyuz, S.; Celik, S. Conformational Analysis and Vibrational Spectroscopic İnvestigation of L-Alanyl-L-Glutamine Dipeptide. *Spectroscopy* **2010**, *24* (3–4), 219–232.

42. Kecel, S.; Ozel, A. E.; Akyuz, S.; Celik, S.; Agaeva, G. Conformational Analysis and Vibrational Spectroscopic İnvestigation of L-Proline–Tyrosine (L-Pro–Tyr) Dipeptide. *J. Mol. Struct.* **2011**, *993* (1), 349–356.

43. Celik, S.; Ozel, A. E.; Akyuz, S.; Kecel, S.; Agaeva, G. Conformational Preferences, Experimental and Theoretical Vibrational Spectra of Cyclo (Gly–Val) Dipeptide. *J. Mol. Struct.* **2011,** *993* (1), 341–348.

44. Celik, S.; Ozel, A. E.; Kecel, S.; Akyuz, S. Structural and IR and Raman Spectral Analysis of Cyclo (His-Phe) Dipeptide. *Vibrat. Spectr.* **2012,** *61,* 54–65.

45. Celik, S.; Ozel, A. E.; Akyuz, S. Comparative Study of Antitumor Active Cyclo (Gly-Leu) Dipeptide: A Computational and Molecular Modeling Study. *Vibrat. Spectr.* **2016,** *83,* 57–69.

46. Celik, S.; Kecel-Gunduz, S.; Akyuz, S.; Ozel, A. E. Structural Analysis, Spectroscopic Characterization and Molecular Docking Studies of the Cyclic Heptapeptide. *J. Biomol. Struct. Dyn.* **2017,** *36* (9), 1–17.

47. Kecel-Gunduz, S.; Bicak, B.; Celik, S.; Akyuz, S.; Ozel, A. E. Structural and Spectroscopic İnvestigation on Antioxidant Dipeptide, ʟ-Methionyl-ʟ-Serine: A Combined Experimental and DFT Study. *J. Mol. Struct.* **2017,** *1137,* 756–770.

48. Celik, S.; Akyuz, S.; Ozel, A. E. Vibrational Spectroscopic and Structural İnvestigations of Bioactive Molecule Glycyl-Tyrosine (Gly-Tyr). *Vibrat. Spectr.* **2017,** *92,* 287–297.

49. Nikiforovich, G. V. Computational Molecular Modeling in Peptide Drug Design. *Chem. Biol. Drug Design* **1994,** *44* (6), 513–531.

50. Han, D.; Wang, H.; Yang, P. Molecular Modeling of Zinc and Copper Binding with Alzheimer's Amyloid β-Peptide. *Biometals* **2008,** *21* (2), 189–196.

51. Fadda, E.; Woods, R. J. Contribution of the Empirical Dispersion Correction on the Conformation of Short Alanine Peptides Obtained by Gas-Phase QM Calculations. *Can. J. Chem.* **2013,** *91* (9), 859–865.

52. Elstner, M.; Jalkanen, K. J.; Knapp-Mohammady, M.; Frauenheim, T.; Suhai, S. Energetics and Structure of Glycine and Alanine Based Model Peptides: Approximate SCC-DFTB, AM1 and PM3 Methods in Comparison with DFT, HF and MP2 Calculations. *Chem. Phys.* **2001,** *263* (2), 203–219.

53. Alvarez, M.; Saavedra, E.; Olivella, M.; Suvire, F.; Zamora, M.; Enriz, R. Theoretical Study of the Conformational Energy Hypersurface of Cyclotrisarcosyl. *Open Chem.* **2012,** *10* (1), 248–255.

54. Beke, T.; Csizmadia, I. G.; Perczel, A. On the Flexibility of β-Peptides. *J. Comput. Chem.* **2004,** *25* (2), 285–307.

55. Bour, P.; Kubelka, J.; Keiderling, T. A. Ab İnitio Quantum Mechanical Models of Peptide Helices and Their Vibrational Spectra. *Biopolymers* **2002,** *65* (1), 45–59.

56. Wiedemann, S.; Metsala, A.; Nolting, D.; Weinkauf, R. The Dipeptide Cyclic (Glycyltryptophanyl) in the Gas Phase: A Concerted Action of Density Functional Calculations, S 0–S 1 Two-Photon Ionization, Spectral UV/UV Hole Burning and Laser Photoelectron Spectroscopy. *Phys. Chem. Chem. Phys.* **2004,** *6* (10), 2641–2649.

57. Marchese, R.; Grandori, R.; Carloni, P.; Raugei, S. On the Zwitterionic Nature of Gas-Phase Peptides and Protein Ions. *PLoS Comput. Biol.* **2010,** *6* (5), e1000775.

58. Kausar, N.; Dines, T. J.; Chowdhry, B. Z.; Alexander, B. D. Vibrational Spectroscopy and DFT Calculations of the Di-Amino Acid Peptide L-Aspartyl-L-Glutamic Acid in the Zwitterionic State. *Phys. Chem. Chem. Phys.* **2009,** *11* (30), 6389–6400.

59. Ghosh, S.; Mondal, S.; Misra, A.; Dalai, S. Investigation on the Structure of Dipeptides: A DFT Study. *J. Mol. Struct.* **2007,** *805* (1), 133–141.

60. Mondal, S.; Chowdhuri, D. S.; Ghosh, S.; Misra, A.; Dalai, S. Conformational Study on Dipeptides Containing Phenylalanine: A DFT Approach. *J. Mol. Struct.* **2007,** *810* (1), 81–89.

61. Kirin, S. I.; Schatzschneider, U.; de Hatten, X.; Weyhermüller, T.; Metzler-Nolte, N. 1,n′-Disubstituted Ferrocenoyl Amino Acids and Dipeptides: Conformational Analysis by CD Spectroscopy, X-Ray Crystallography, and DFT Calculations. *J. Organomet. Chem.* **2006,** *691* (16), 3451–3457.

62. Testa, B.; Kier, L. B. The Concept of Molecular Structure in Structure–Activity Relationship Studies and Drug Design. *Med. Res. Rev.* **1991,** *11* (1), 35–48.

63. Peterson, L. R. Quinolone Molecular Structure–Activity Relationships: What We Have Learned About İmproving Antimicrobial Activity. *Clin. Infect. Dis.* **2001,** *33* (Supplement 3), S180–S186.

64. Kubinyi, H. Quantitative Structure–Activity Relationships in Drug Design. *Encycl. Comput. Chem.* **1998,** *4*, 2309–2320.

65. Kandemirli, F. *Introductory Chapter: Some Quantitative Structure Activity Relationship Descriptor, Quantitative Structure–Activity Relationship*; Kandemirli, F., Ed.; InTech, 2017. doi:10.5772/intechopen.69642.

66. Tillotson, G. S. Quinolones: Structure–Activity Relationships and Future Predictions. *J. Med. Microbiol.* **1996,** *44* (5), 320–324.

67. Kier, L. B.; Hall, L. H. Molecular Connectivity in Structure–Activity Analysis. *Res. Stud.* **1986.**

68. Topliss, J. (Ed.). *Quantitative Structure–Activity Relationships of Drugs* (Vol. 19); Elsevier, 2012.

69. Bowers, A. A.; Greshock, T. J.; West, N.; Estiu, G.; Schreiber, S. L.; Wiest, O.; … Bradner, J. E. Synthesis and Conformation–Activity Relationships of the Peptide Isosteres of FK228 and Largazole. *J. Am. Chem. Soc.* **2009,** *131* (8), 2900–2905.

70. Ramachandhan, G. N. Need for Nonplanar Peptide Units in Polypeptide Chains. *Biopolymers* **1968,** *6* (10), 1494–1496.

71. Ramachandran, G. N.; Ramakrishnan, C.; Sasisekharan, V. Stereochemistry of Polypeptide Chain Configurations. *J. Mol. Biol.* **1963,** *7* (1), 95–99.

72. Maksumov, I. S.; Ismailova, L. I.; Godzhaev, N. M. A Program for the Semiempirical Calculation of the Conformations of Molecular Complexes on a Computer. *J. Struct. Chem.* **1984,** *24* (4), 647–648.

73. Scott, R. A.; Scheraga, H. A. Conformational Analysis of Macromolecules. III. Helical Structures of Polyglycine and Poly-L-Alanine. *J. Chem. Phys.* **1966,** *45* (6), 2091–2101.

74. Nemethy, G.; Pottle, M. S.; Scheraga, H. A. Energy Parameters in Polypeptides. 9. Updating of Geometrical Parameters, Nonbonded İnteractions, and Hydrogen Bond İnteractions for the Naturally Occurring Amino Acids. *J. Phys. Chem.* **1983,** *87* (11), 1883–1887.

75. Momany, F. A.; McGuire, R. F.; Burgess, A. W.; Scheraga, H. A. Energy Parameters in Polypeptides. VII. Geometric Parameters, Partial Atomic Charges, Nonbonded İnteractions, Hydrogen Bond İnteractions, and İntrinsic Torsional Potentials for the Naturally Occurring Amino Acids. *J. Phys. Chem.* **1975,** *79* (22), 2361–2381.

76. Scheraga, H. A. Calculations of Conformations of Polypeptides. *Adv. Phys. Org. Chem.* **1968,** *6*, 103–184.

77. Celik, S.; Özel, A. In *Structural Analysis of Biologically Active Tetrapeptide*, 9th International Physics Conference of the Balkan Physical Union, Istanbul, Turkey, Aug 24–27, 2015; p 354.

78. Agaeva, G. A.; Hasanova, N. G.; Godjaev, N. M. In *Structural İnformation About Antihypertensive Peptide Ovokinin (2–7), Obtained by Computer Modeling*, Application of Information and Communication Technologies (AICT), 2011 5th International Conference on. IEEE, 2011; pp 1–3.

79. Agaeva, G. A.; Agaeva, U. T.; Godjaev, N. M. The Spatial Organization of the Human Hemokinin-1 and Mouse/Rat Hemokinin-1 Molecules. *Biophysics* **2015,** *60* (3), 365–377.

80. Agaeva, G. A. Conformational Spatial Structure of the Tachykinin Peptide Eledoisin. *J. Qafqaz Univ.* **2010,** *29,* 17–22.

81. Metropolis, N. A.; Rosenbluth, A. W.; Rosenbluth, M. N.; Teller, A. H.; Teller, E. Equation of State Calculations by Fast Computing Machines. *J. Chem. Phys.* **1953,** *21,* 1087–1092.

82. Rahman, A. Correlations in the Motion of Atoms in Liquid Argon. *Phys. Rev.* **1964,** *136,* A405–A411.

83. Mehrazma, B.; Robinson, M.; Opare, S. K. A.; Petoyan, A.; Lou, J.; Hane, F. T.; … Leonenko, Z. Pseudo-Peptide Amyloid-β Blocking İnhibitors: Molecular Dynamics and Single Molecule Force Spectroscopy Study. *Biochim. Biophys. Acta Proteins Proteom.* **2017,** *1865* (11), 1707–1718.

84. Istrate, A. N.; Kozin, S. A.; Zhokhov, S. S.; Mantsyzov, A. B.; Kechko, O. I.; Pastore, A.; … Polshakov, V. I. Interplay of Histidine Residues of the Alzheimer's Disease Aβ Peptide Governs İts Zn-Induced Oligomerization. *Sci. Rep.* **2016,** *6,* 21734.

85. Kecel-Gündüz, S.; Budama-Kilinc, Y.; Cakir Koc, R.; Kökcü, Y.; Bicak, B.; Aslan, B.; Özel, A. E. Computational Design of Phe-Tyr Dipeptide and Preparation, Characterization, Cytotoxicity Studies of Phe-Tyr Dipeptide Loaded PLGA Nanoparticles for the Treatment of Hypertension. *J. Biomol. Struct. Dyn.* **2017,** *36* (11), 1–15.

86. DeVane, R.; Ridley, C.; Larsen, R. W.; Space, B.; Moore, P. B.; Chan, S. I. A Molecular Dynamics Method for Calculating Molecular Volume Changes Appropriate for Biomolecular Simulation. *Biophys. J.* **2003,** *85,* 2801–2807. doi:10.1016/S0006-*3495* (03)74703-1.

87. Scheraga, H. A.; Khalili, M.; Liwo, A. Protein-Folding Dynamics: Overview of Molecular Simulation Techniques. *Ann. Rev. Phys. Chem.* **2007,** *58,* 57–83. doi:10.1146/annurev.physchem.58.032806.104614.

88. Komaromi, I.; Somogyi, A.; Wysocki, V. H. Proton Migration and İts Effect on the MS Fragmentation of N-Acetyl OMe Proline: MS/MS Experiments and AB Initio and Density Functional Calculations. *Int. J. Mass Spectr.* **2005,** *241* (2), 315–323.

89. Yao, G.; Zhang, J.; Huang, Q. Conformational and Vibrational Analyses of Meta-Tyrosine: An Experimental and Theoretical Study. *Spectrochim. Acta A Mol. Biomol. Spectrosc.* **2015,** *151,* 111–123.

90. Jorgensen, W. L. Rusting of the Lock and Key Model for Protein-Ligand Binding. *Science* **1991,** *254* (5034), 954–956.

91. Wei, B. Q.; Weaver, L. H.; Ferrari, A. M.; Matthews, B. W.; Shoichet, B. K. Testing a Flexible-Receptor Docking Algorithm in a Model Binding Site. *J. Mol. Biol.* **2004,** *337* (5), 1161–1182.

92. Lengauer, T.; Rarey, M. Computational Methods for Biomolecular Docking. *Curr. Opin. Struct. Biol.* **1996,** *6* (3), 402–406.

93. Trott, O.; Olson, A. J. AutoDock Vina: Improving the Speed and Accuracy of Docking with a New Scoring Function, Efficient Optimization, and Multithreading. *J. Comput. Chem.* **2010,** *31* (2), 455–461.

94. Goldman, B. B.; Wipke, W. T. QSD Quadratic Shape Descriptors. 2. Molecular Docking Using Quadratic Shape Descriptors (QSDock). *Proteins* **2000,** *38* (1), 79–94.

95. Meng, E. C.; Shoichet, B. K.; Kuntz, I. D. Automated Docking with Grid-Based Energy Evaluation. *J. Comput. Chem.* **2004,** *13* (4), 505–524.

96. Feig, M.; Onufriev, A.; Lee, M. S.; Im, W.; Case, D. A.; Brooks, C. L. Performance Comparison of Generalized Born and Poisson Methods in the Calculation of Electrostatic Solvation Energies for Protein Structures. *J. Comput. Chem.* **2004,** *25* (2), 265–284.

97. Dhanavade, M. J.; Parulekar, R. S.; Kamble, S. A.; Sonawane, K. D. Molecular Modeling Approach to Explore the Role of Cathepsin B from *Hordeum vulgare* in the Degradation of Aβ Peptides. *Mol. BioSyst.* **2016,** *12* (1), 162–168.

98. Meng, X. Y.; Zhang, H. X.; Mezei, M.; Cui, M. Molecular Docking: A Powerful Approach for Structure-Based Drug Discovery. *Curr. Comput. Aided Drug Des.* **2011,** *7,* 146–157. doi:10.2174/157340911795677602.

99. Weinstock, J.; Keenan, R. M.; Samanen, J.; Hempel, J.; Finkelstein, J. A.; Franz, R. G.;…Gleason, J. G. 1-(Carboxybenzyl) Imidazole-5-Acrylic Acids: Potent and Selective Angiotensin II Receptor Antagonists. *J. Med. Chem.* **1991,** *34* (4), 1514–1517.

100. Keenan, R. M.; Weinstock, J.; Finkelstein, J. A.; Franz, R. G.; Gaitanopoulos, D. E.; Girard, G. R.; … Samanen, J. M. Potent Nonpeptide Angiotensin II Receptor Antagonists. 2. 1-(Carboxybenzyl) İmidazole-5-Acrylic Acids. *J. Med. Chem.* 1993, *36* (13), 1880–1892.

101. Buckingham, J.; Glen, R. C.; Hill, A. P.; Hyde, R. M.; Martin, G. R.; Robertson, A. D.; … Woollard, P. M. Computer-Aided Design and Synthesis of 5-Substituted Tryptamines and Their Pharmacology at the 5-HT1D Receptor: Discovery of Compounds with Potential Anti-Migraine Properties. *J. Med. Chem.* 1995, *38* (18), 3566–3580.

102. Dorsey, B. D.; Levin, R. B.; McDaniel, S. L.; Vacca, J. P.; Guare, J. P.; Darke, P. L.; Zugay, J. A.; Emini, E. A.; Schleif, W. A. L-735,524: The Design of a Potent and Orally Bioavailable HIV Protease Inhibitor. *J. Med. Chem.* 1994, *37* (21), 3443–3451.

103. Holloway, M. K.; et al. In *Computer-Aided Molecular Design*; Reynolds, C. H., et al., Eds., ACS Symp. Series 589, 1995; pp 36–50.

104. Greer, J.; Erickson, J. W.; Baldwin, J. J.; Varney, M. D. Application of the Three-Dimensional Structures of Protein Target Molecules in Structure-Based Drug Design. *J. Med. Chem.* 1994, *37* (8), 1035–1054.

105. Baldwin, J. J.; Ponticello, G. S.; Anderson, P. S.; Christy, M. E.; Murcko, M. A.; Randall, W. C.; Schwam, H.; Sugrue, M. F.; Gautheron, P. Thienothiopyran-2-Sulfonamides: Novel Topically Active Carbonic Anhydrase İnhibitors for the Treatment of Glaucoma. *J. Med. Chem.* 1989, *32* (12), 2510–2513.

106. Li, G. Theoretical Studies of Anti-Cancer Drug Tamoxifen and Estrogen Receptor Alpha. Doctoral Dissertation, KTH Royal Institute of Technology, 2012.

107. Rungnim, C.; Arsawang, U.; Rungrotmongkol, T.; Hannongbua, S. Molecular Dynamics Properties of Varying Amounts of the Anticancer Drug Gemcitabine

İnside an Open-Ended Single-Walled Carbon Nanotube. *Chem. Phys. Lett.* 2012, *550*, 99–103.

108. Yang, X.; Wang, Z.; Xiang, Z.; Li, D.; Hu, Z.; Cui, W.; … Fang, Q. Peptide Probes Derived from Pertuzumab by Molecular Dynamics Modeling for HER2 Positive Tumor İmaging. *PLoS Comput. Biol.* 2017, *13* (4), e1005441.

109. Otuokere, I. E.; Amaku, F. J. Molecular Mechanics Geometry Optimization and Excited-State Properties of Cardioprotective Drug 4,4′-(2S)-Propane-1,2-Diyldipiperazine-2,6-Dione (Dexrazoxane). *Res. J. Pharmacol. Pharmacodyn.* 2015, *7* (3), 137–142.

110. Ekins, S.; et al. *Pharmacophore Perception, Development and Use in Drug Design*; Güner, O. F., Ed., La Jolla, 2000; pp 269–288.

111. Sachdeva, R.; Singh, V. P.; Saini, G. S. S. Vibrational Spectroscopic Studies of Sildenafil in 1800–900 cm^{-1} Region. *AIP Conf. Proc.* 2015, *1675* (1), 030075.

CHAPTER 7

A Sustainable, Efficient, and Green Promoter for the Synthesis of Some Heterocyclic Compounds

SANGEETA BHARGAVA*, ANITA CHOUDHARY, and DEEPTI RATHORE

Department of Chemistry, Centre of Advanced Studies, University of Rajasthan, Jaipur 302004, India

Corresponding author. E-mail: drsbhargava1@gmail.com

ABSTRACT

Imidazolium based ionic liquids were used for the synthesis of a diverse library of benzo[a]phenothiazines. Ionic liquids were synthesized and used as solvents as well as catalyst. This reaction offered many advantages including simplicity of operation, short reaction time, high product yield, recyclability and eco-friendly nature of the reaction medium. The structures of all the synthesized compounds were established by IR, ^{1}H-NMR, ^{13}C-NMR and elemental analysis.

7.1 INTRODUCTION

Heterocyclic chemistry is one of the most intricate and fascinating branches of organic chemistry that deals with the synthesis, properties, and applications of heterocyclic compounds.[1–3] As defined by IUPAC, heterocyclic compounds are "cyclic compounds having as ring member, atoms of at least two different elements." The suffix cyclic in heterocyclic indicates that at least one ring is present in the compound and prefix hetero refers to the heteroatoms present in this ring. The heteroatoms commonly found in these compounds are oxygen, nitrogen, and sulfur. According to the heteroatom(s) present in the ring, these compounds can be classified as oxygen, nitrogen, or sulfur based. Compounds are also organized based on the size of the ring that is determined by the total number of atoms present in the ring. The size and type of ring structures, along with the substituent groups attached to the core moiety, strongly impact the physicochemical properties of these compounds.[4-6]

More than half of the known organic compounds belong to the category of heterocycles and nonaromatic heterocycles are more abundant than hetero-aromatics.[7] They are inevitably woven into our lives due to their widespread occurrence in natural as well as synthetic products. They form the cornerstone of pharmaceutical and agrochemical industries. Moreover, they constitute the core moiety of several biologically active compounds, including antitumor, antibiotic, anti-inflammatory, antidepressant, antimalarial, anti-HIV, antimi-crobial, antibacterial, antifungal, antiviral, and antidiabetic agents.[8-11] Almost all the drugs that exist in the market today contain heterocycles.

In addition to the rising levels of pollution in the environment, it has become imperative for the scientists to design and develop such methodologies for the synthesis of heterocyclic compounds that abide by the principles of green chemistry. The 12 principles of green chemistry can be considered as the blueprint for chemists to accomplish the goal of sustainable development. They constitute an overarching construct for the design of safer chemicals and chemical transformations. In today's world, chemicals form a vital role in our day-to-day life. They are used in a wide range of products and processes, and their sound management is essential in order to avoid substantial and increasingly complex risks to human health and ecosystem. This is one of the biggest challenges that chemistry has to face. So, the pressing need for the synthetic chemists is to minimize chemical pollution by following the green protocol. One way to do so is by designing and developing such solvents that are environmentally benign. It has been recognized that employing nonconventional solvents as alternatives to environmentally hazardous traditional solvents for organic synthesis can reduce waste solvent production and hence reduce the harmful effect on environment to a large extent. Some of the newly devised eco-friendly solvent systems are water, supercritical fluids (such as supercritical CO_2), fluorinated solvents, ionic liquids, and solventless processes.[12-15]

Of the aforementioned solvents, ionic liquids have emerged as a highly promising alternative to the conventional organic solvents. In recent years, interest in ionic liquids has burgeoned due to their rising applications in the field of research and technology. Ionic liquids may be defined as organic salts that are entirely composed of ions and are found in molten state below 100 °C.[16] They have been termed as "green solvents" due to their unique physiochemical properties such as nonflammability, nonvolatility, recyclability, thermal stability, wide electrochemical window, viscosity, density, refractive index, and hydrophobicity[17–23] (Fig. 7.1).

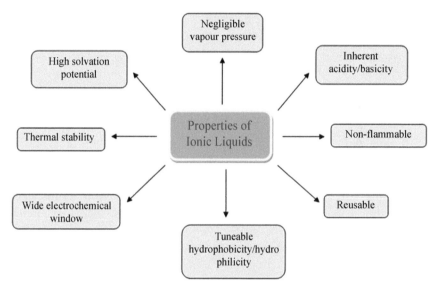

FIGURE 7.1   Properties of ionic liquids.

Another important feature of ILs is that their properties can be subtly tailored to suit the reaction condition by varying the combination of cations and anions or changing the substituents on cation or anion.[24] This ability to fine-tune their physical properties has earned them the accolade of designer solvents.[25] Furthermore, their high solvation potential is able to generate an internal pressure and promote the association of reactants in the solvent cavity leading to higher reaction rates.[26] Also, product separation from ionic liquids is more convenient than conventional solvents. These advantages make them the solvent of choice for a large number of chemical processes[27–30] (Fig. 7.2). Even though ionic liquids had initially been introduced as green alternatives to conventional volatile organic solvents, they have now marched far beyond this boundary, playing a prominent role as catalyst, gas adsorbent, and chromatography stationary phases, etc.[31–34]

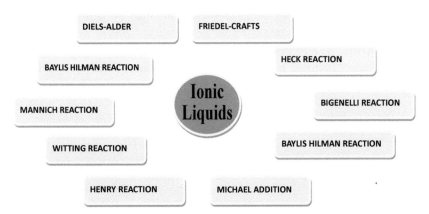

FIGURE 7.2   Applications of ionic liquid.

In line with the growing applications of ionic liquids in synthetic organic chemistry, we have in our work presented here prepared a series of benzo[*a*]phenothiazine derivatives using ionic liquid as a green and eco-compatible solvent. Imidazolium-based ionic liquids were synthesized and used as solvents. Further, the reactivity of the recycled ionic liquid was also studied and all the synthesized compounds were characterized using various spectroscopic techniques.

7.1.1   BRIEF OVERVIEW OF PHENOTHIAZINES

Phenothiazine constitutes one of the most widely researched class of thiazine heterocyclic ring system containing both nitrogen and sulfur heteroatoms. Over the years, studies on phenothiazine derivatives have gained wide acceptability and different derivatives of this heterocyclic system both linear of the type (**I**) and nonlinear of the type (**II**) analogs have been prepared and reported.[35-40]

I   II

The earliest recorded synthesis of phenothiazine compounds **(III)** that act as purple dyes was done in 1876 by Lauth.[41] Caro also synthesized methylene blue dyes containing phenothiazine nucleus **(IV)**.

III	**IV**

The structures of the two dyes were unknown until Berthensen[42] revealed that they were the derivatives of phenothiazine by the synthesis of phenothiazine **(V)** in 1883.

V

A family of these heterocycles, including the Benzo[*a*], Benzo[*b*], Benzo[*c*] phenothiazine and their derivatives, have been synthesized in several ways.[8–13] Many bioactive heterocyclic compounds, which have wide use in industrial, biological, and pharmacy region, contain phenothiazine nucleus.[43–54] They possess various pharmacological and therapeutical properties such as antiallergic,[55] antileukemic,[56,57] antimutagenic,[58,59] anti-inflammatory,[60] antibacterial,[61–63] antifungal,[64,65] antiviral,[66–68] antiplasmid,[69] antimalarial,[70] anticarcinogenic,[71–73] radical modulation,[74] antitumor,[75–82] cytotoxic,[83–85] immunomodulator,[86] agonist,[87] multidrug resistance (MDR) and tumor antigen expression,[88,89] antitubercular activity,[90] cholinesterase inhibitor,[91] histamine antagonist,[92] and neuromuscular incoordination agent.[93]

They are used as dyes and pigments in textile, paint, and plastic industries[94–101] and as insecticides in agricultural industries.[102] They are also found useful as antioxidants in lubricants and fuels in petroleum industries.[103–108]

Foley and Cincotta[109] synthesized 5-imino-9-diethylamino-1,2,3,4-tetrahydrobenzo[*a*]phenothiazine (**1**) by stirring a mixture of 4,5,7,8-tetra-hydro-1-naphthylamine, 2-amino-5-dimethylaminophenyl thiosulfuric acid, potassium chromate, and dimethyl sulfoxide.

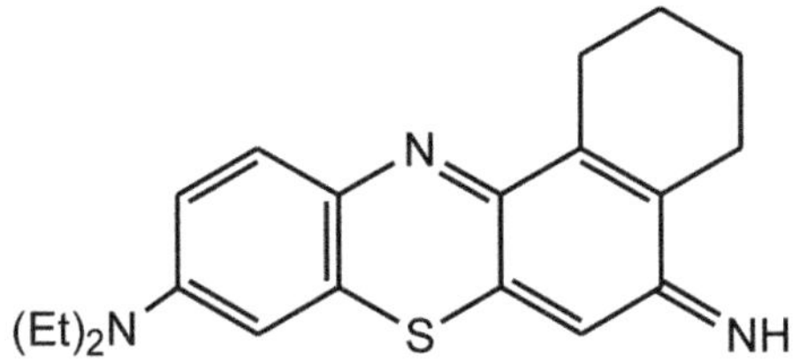

1

Motohashi et al.[110] studied the relationship between πe⁻ distribution and anticancer activity of 9-methyl-12*H*-benzo[*a*]phenothiazine (**2**) and demonstrated that the stability of π electrons in benzene ring and naphtha-lene ring in the benzo[*a*]phenothiazine induces anticancer activity.

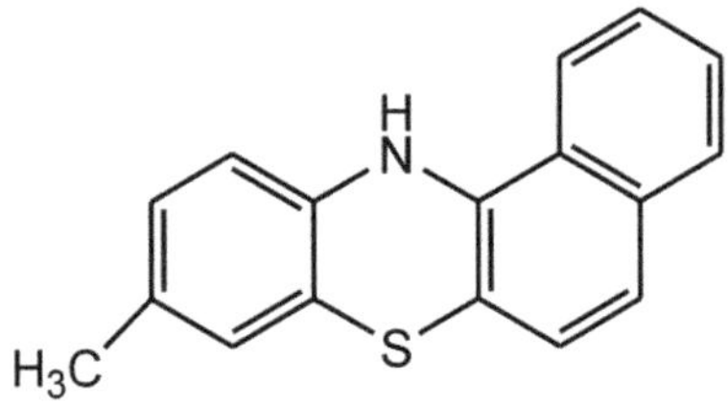

2

The photochemotherapeutic properties of a novel benzophenothiazine (**4**), 5-ethylamino-9-diethylaminobenzo[*a*]phenothiazinium chloride (**3**) were assessed in vitro and in vivo against two murine mammary sarcoma models (EMT-6 and RIF) by Cincotta et al.[111]

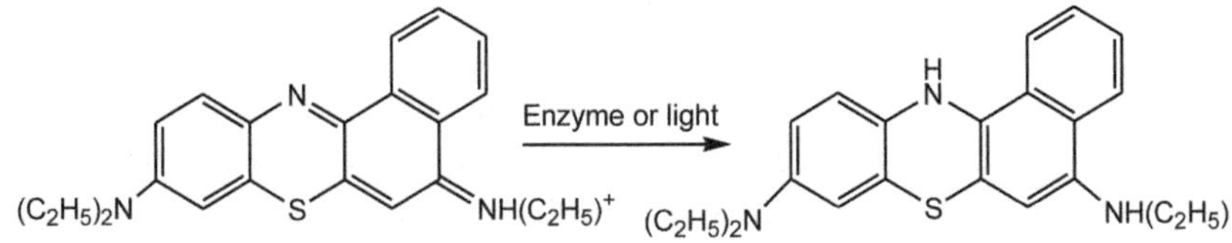

3   **4**

Prakash et al.[112] prepared substituted benzo[*a*]phenothiazines (**5**) and (**6**) as antimicrobial agents.

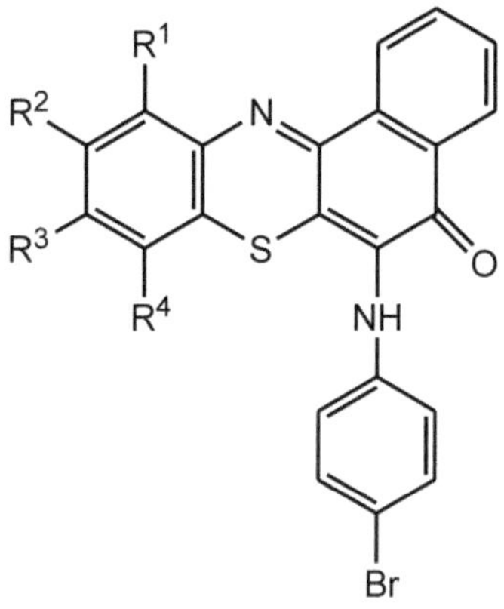

5 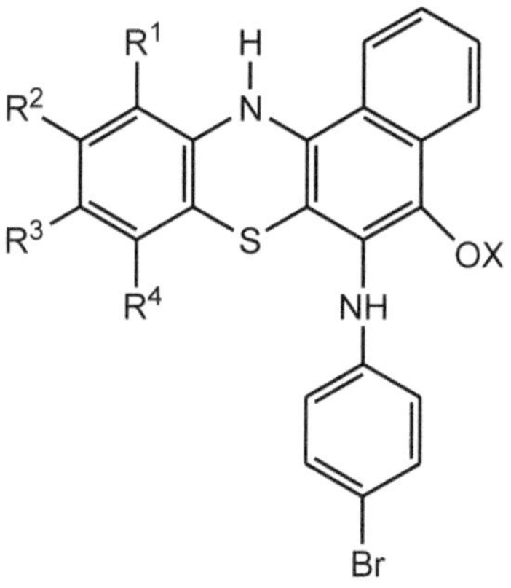   **6**

where X = H, Ac, CH₃

where X = H, Ac, CH$_3$

Aaron et al.[113] examined solvatochromic study of benzo[*a*]phenothiazine (**7**) for the calculation of dipole moment and specific solute–solvent interaction in the first excited singlet state.

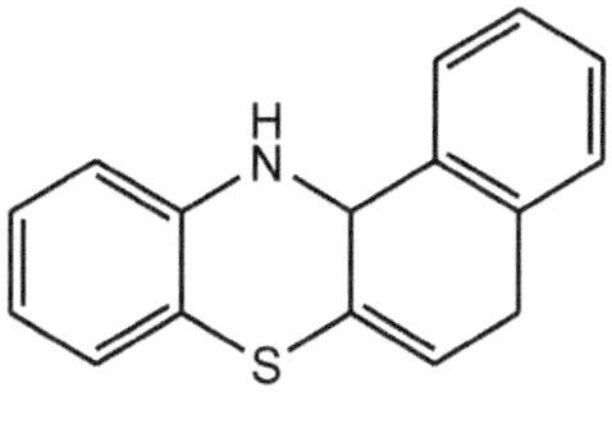

7

Pusztai et al.[114] investigated effect of benzo[*a*]phenothiazines and its derivatives on cancerous cells and showed that the presence of the methyl group increased the T-antigen activity at higher dose concentrations, while the hydroxy substituent decreases the T-antigen activity. It has also been demonstrated that the methyl substitution at positions C-9 or C-10 has enhanced the T-antigen expression of adenovirus infected cells.

Kurihara et al.[115] studied the relationship between radical production and π spin density of benzo[*a*]phenothiazine (**7**) by UHF/PM3 method and found that π spin density is positive in radical generating benzo[*a*]

phenothiazine and nonradical generating benzo-phenothiazine had opposite π spin density.

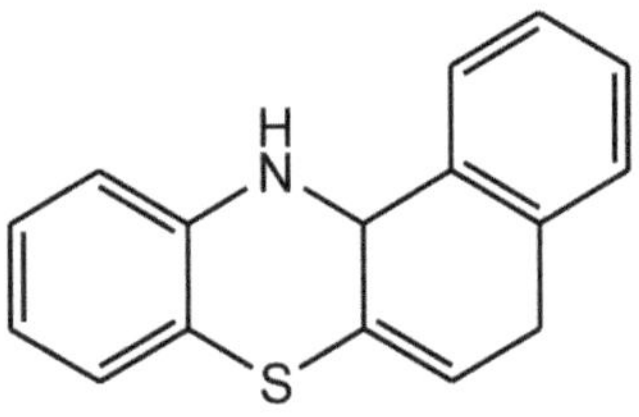

7

- Ribofuranosides **(9)** of fluoro-substituted-benzo[*a*]phenothiazines
  (8) have been synthesized by condensation of phenothiazines with
  b-D-ribofuranose-1-acetate-2,3,5-tribenzoate by Swati et al.[116]
  These derivatives have shown antimicrobial activity.

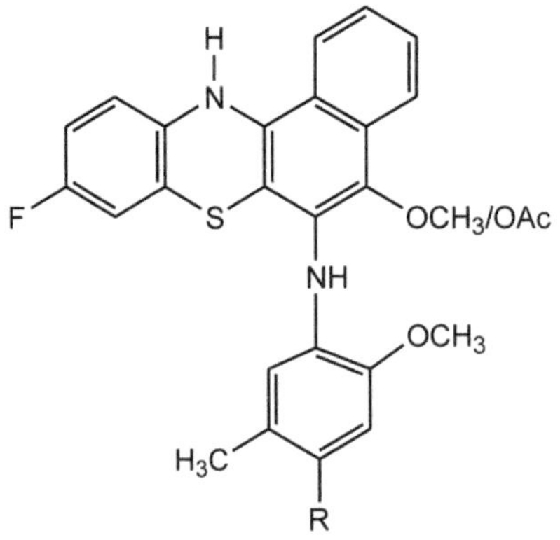

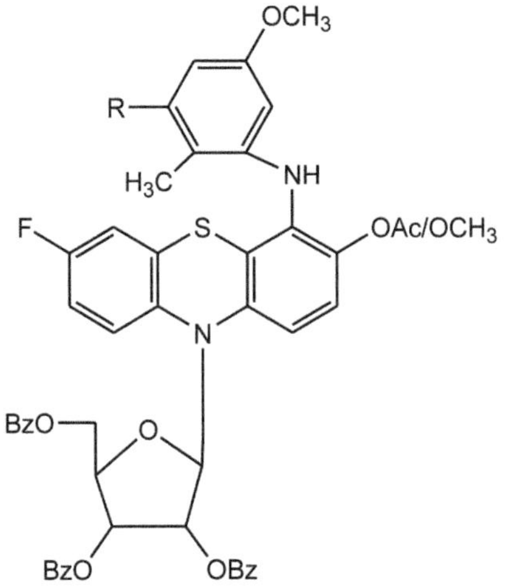

8　　　　　　　　　　　　　**9**

where R = H, Cl

Mucsi et al.[117] examined combined antiviral effects of some benzo[*a*]
phenothiazines **(10)**, **(11)** and 9-[2-hydroxy(ethoxy)methyl] guanine
(acycloguanosine, acyclovir, ACV) on the multiplication of herpes simplex
virus type 2 (HSV-2) using Vero cells.

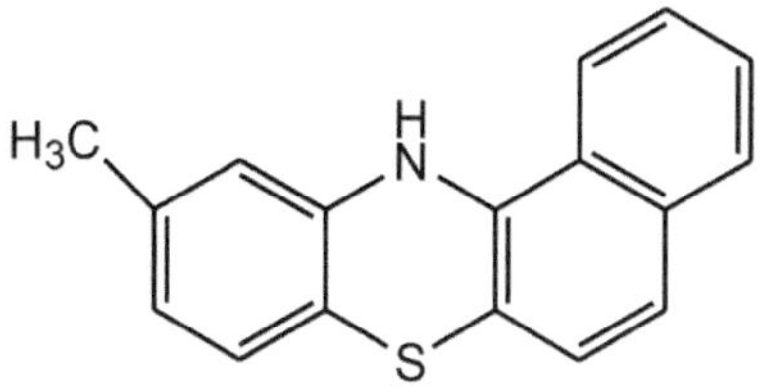

10 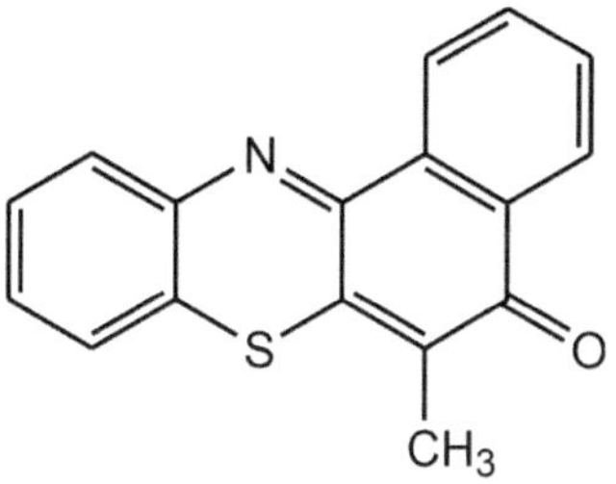 **11**

Ektova et al.[118] presented an efficient synthesis of 4-hydroxy-5-*H*-benzo[*a*]phenothiazine-5-imines **(14)** by the reaction of zinc salts of 2-aminobenzenethiols **(12)** and *N*-aryl-2,6-dibromo/2,6,8-tribromo-5-hydroxy-1,4-naphthoquinone-4-imines **(13)**.

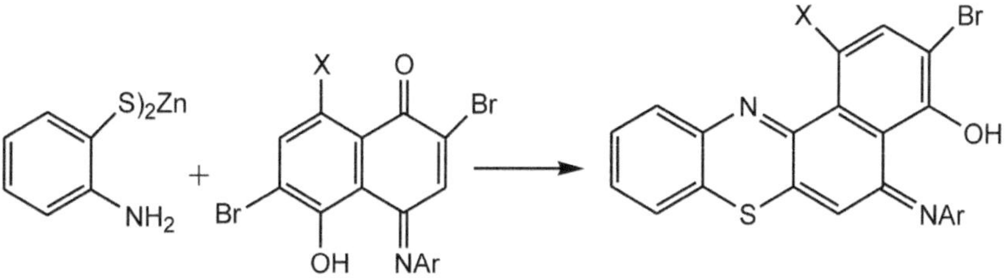

12 **13** **14**

Ar = Ph, C$_6$H$_4$-OC$_4$H$_9$

X = H, Br

Hilgeroth et al.[119] synthesized 5-oxo-5*H*-benzo[*a*]phenothiazine **(15)** and 6-hydroxy-5-oxo-5*H*-benzo[*a*]phenothiazine **(16)** and evaluated the relation between calculated molecular orbital energies with MDR properties.

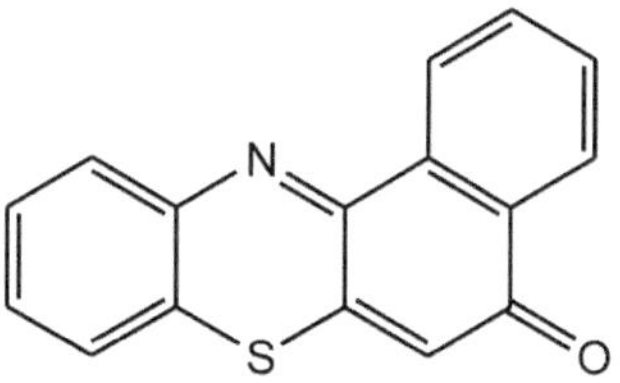

15 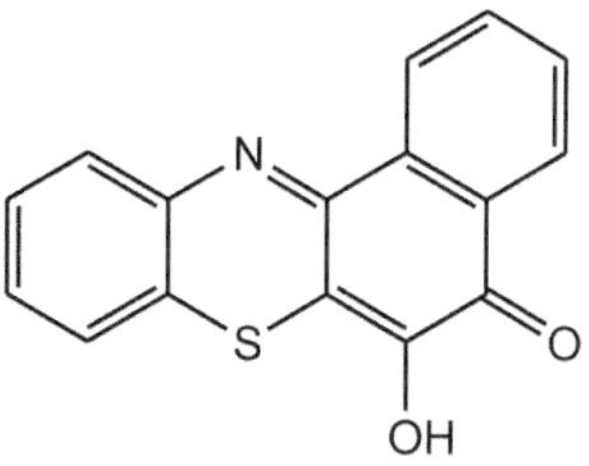 **16**

5*H*-benzo[*a*]phenothiazine (**17**) and 12*H*-benzo[*a*]phenothiazine (**18**) have been synthesized by reaction of zinc mercaptide of substituted amino-benzenethiol and 2-(4-chloro-2-methoxy-5-methylanilino/2-methoxy-5-methylanilino)-3-chloro-1,4-naphthoquinone by Yadav et al.[120]

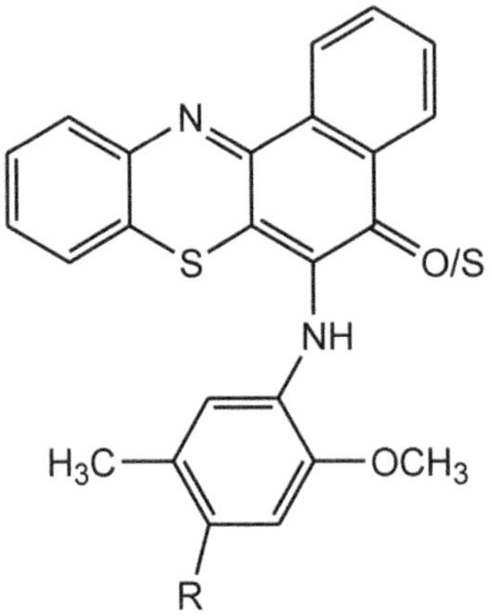

17 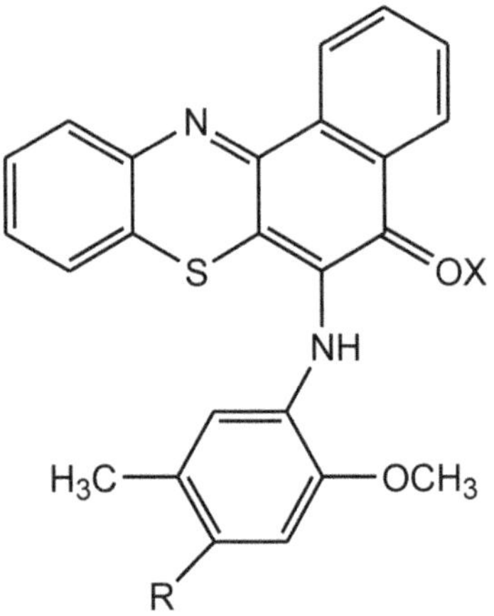   **18**

X = H, COCH₃, CH₃

Quantitative structure–activity relationship analysis for minimum inhibitory concentration of phenothiazines and benzo[*a*]phenothiazines (**19**) was investigated by Kurihara et al.[121]

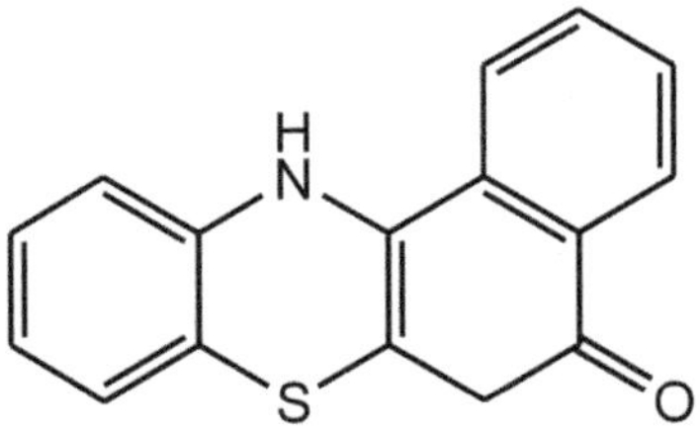

19

Kang et al.[122] have synthesized benzo[*a*]phenothiazine derivatives (**21**) by the reaction of dichloronaphthoquinones (**20**) with 2-aminothiophenol in EtOH-HCl.

20   **21**

Aaron et al.[123] reported the spectroscopic studies, biological and biomedical properties, and applications of benzo[*a*]phenothiazines. 6-acetamido benzo[*a*] phenothiazin-5-one **(24)** has been synthesized by reaction between condensed 2-aminobenzenethiol **(22)** and 3-acetamido-2-chloro-1,4-naphthoquinone **(23)** in the presence of alcohol.[124]

22   **23**   **24**

2,3-dichloro-1,4-naphthoquinone **(25)** reacts with sulfur nucleophile **(26)** resulting in the formation of 2-arylsulfanyl-3-chloro-[1,4] naphtho-quinones **(27)** which produced 12*H*-benzo[*b*]phenothiazine-6,11-diones **(28)** and their analogs on nucleophilic displacement reaction with NaN_3 in DMF–H_2O (10:1) by Tandon et al.[125]

25   **27**

6-Alkylbenzo[*a*]phenothiazin-5-one **(31)** has been synthesized via nickel-catalyzed Grignard couplings by Ezema et al.[126]

The synthesis of 6-aryl derivatives of 8-methyl-11-azabenzo[*a*] phenothiazn-5-one **(35)** was by Ijeomah et al.[127] This was achieved by reacting 2-amino-4-methylpyridine-3-thiol **(32)** with 2,3-dichloro-1-4-naphthoquinone **(33)** under anhydrous basic condition.

Ezema et al.[128] reported synthesis of isomeric derivatives of angular aza benzophenothiazine **(40)** **(41)** from the reactions of 2-amino-3,5-dibromopyrazine **(36)** with 6,7-dibromoquinoline-5,8-dione **(37)** in anhydrous sodium carbonate. **(38)** **(39)** compounds further react with 4-aminopyrimidin-5-thiol to form azabenzothiazinophenothiazine rings **(40)** **(41)**.

Ayuk et al.[129] reported the synthesis of 6-(4-bromophenyl)-10-methyl-11-azabenzo[*a*]phenothiazin-5-one **(44)**. It was prepared via transition metal-catalyzed cross-coupling reaction.

Fu et al.[130] used molecular hybridization technique to design new phenothiazine–dithiocarbamate derivatives **(45)** and also studied their anticancer activity.

Hu et al.[131] developed an iron-catalyzed synthetic method for the synthesis of phenothiazines **(47)** via a tandem C–S/C–N cross-coupling reaction.

Inspired by the various biological activities of benzo[*a*]phenothiazines, a simple, green, and eco-friendly novel protocol has been developed for their synthesis by the tandem reaction of readily available reagents in imidazolium-based ionic liquids. The ionic liquid has been used as a solvent as well as catalyst for this reaction. This reaction offered many advantages, including simplicity of operation, short reaction time, high yield of product, recyclability, and eco-friendly nature of the reaction medium. The structures of all the synthesized compounds have been established by IR, ^{1}H-NMR, ^{13}C-NMR, and elemental analysis.

7.2 EXPERIMENTAL

7.2.1 GENERAL

All the chemicals used were of research grade (purchased from Sigma Aldrich, Merck, and Acros). Melting points of all the synthesized compounds were determined in open capillary tubes and are uncorrected. Purity of compounds was checked on thin layers of silica Gel-G coated glass plates and *n*-hexane:ethyl acetate (8:2) as eluent and visualization was accomplished by iodine or UV light. IR spectra were recorded on a SHIMADZU FT-IR spectrometer using KBr Pellets. ^{1}H NMR spectra were recorded on a JEOL-AL-300 MHz NMR spectrometer in DMSO and CDCl$_3$ using TMS as an internal standard (chemical shift in δ ppm). ^{13}C NMR spectra were recorded on a JEOL-AL-75 MHz NMR spectrometer in DMSO and CDCl$_3$ using TMS as an internal standard.

7.2.2 SYNTHESIS OF IONIC LIQUIDS

In order to exploit the potential of ionic liquids for the synthesis of hetero-cyclic compounds, imidazolium-based ionic liquids were first prepared according to Scheme 7.1:

7.2.2.1 SYNTHESIS OF 1-METHYLIMIDAZOLE

34 g (0.5 mmol) of imidazole was taken in a round-bottom flask and dissolved in 30 mL dried methanol. A freshly prepared solution of

sodium methoxide (0.55 mol) was added to the above mixture. The resulting solution was magnetically stirred for 30 min. Then 0.55 mol of alkyl bromide was added dropwise over a period of 1 h. The contents of the flask were refluxed for 30 min, cooled to room temperature, and filtered to remove sodium bromide. The filtrate was concentrated under reduced pressure at 70 °C and later filtered to remove undissolved sodium bromide. The filtrate thus obtained was distilled to afford 1-alkylimidazole.

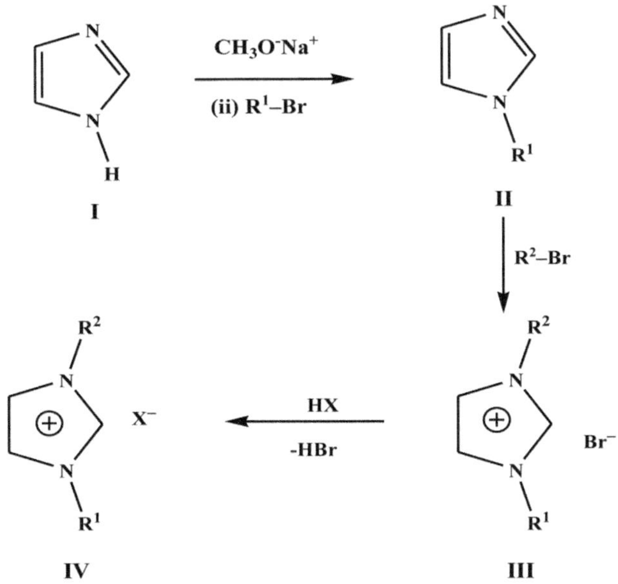

R_1 = -CH$_3$, -C$_2$H$_5$, -C$_4$H$_9$;
R_2 = -CH$_3$, -C$_4$H$_9$;
X = BF$_4$-, PF$_6$-, -CH$_3$SO$_3$-,
-CF$_3$COO-, -CF$_3$SO$_3$-, HSO$_4$-

SCHEME 7.1   Synthesis of dialkylated imidazolium based ionic liquid.

7.2.2.2   SYNTHESIS OF 1-BUTYL-3-METHYLIMIDAZOLIUM BROMIDE

A mixture of 1-methylimidazole (0.1 mol) and *n*-butylbromide (0.1 mol) was mixed in a round-bottom flask and refluxed for 48 h at 70 °C with stirring until the formation of two phases. The top layer containing unreacted material was decanted and washing was done thrice using ethyl acetate. The remaining solvent was removed on rotavapor to obtain the desired ionic liquid.

7.2.2.3 SYNTHESIS OF 1,3-DIBUTYLIMIDAZOLIUM AND 1-BUTYL-3-METHYLIMIDAZOLIUM IONIC LIQUIDS [ANION]—, BF_4^-, PF_6^-

A solution of [BMIM]Br (0.30 mol) in dichloromethane was taken in a round-bottom flask and the respective metal salt ($NaBF_4$/$NaPF_6$) (0.32 mol) required to form the desired ionic liquid was added to it and stirred magnetically for 28 h. The suspension was filtered to remove the precipitated bromide salt and the organic phase was washed several times with water until no precipitation of AgBr occurred in the aqueous phase on adding a concentrated $AgNO_3$ solution. The organic phase was again washed thrice with water to guarantee complete removal of the bromide salt. The solvent was then removed on rotavapor to give a slightly yellowish ionic liquid.

Using the above-synthesized ionic liquids, we then prepared benzo[*a*] phenothiazine derivatives as per the following steps.

7.2.3 SYNTHESIS OF 2-AMINOBENZOTHIAZOLES (3A–F)

A solution of substituted aniline (0.4 mmol) **1a–f** and ammonium thiocyanate (0.4 mmol) **2** in glacial acetic acid was cooled in ice bath and stirred mechanically. To the above stirring reaction, content was added a previously cooled solution of bromine (5.3 mL) in glacial acetic acid (25 mL) in dropwise manner. The reaction mixture was stirred further for 2 h and temperature was maintained below 10 °C. The reaction contents were then left overnight in a refrigerator. The precipitate of benzene–thiazole hydrobromide was dissolved in warm water, filtered and then the filtrate was neutralized with aqueous solution of Na_2CO_3. The precipitate obtained on neutralization was collected by filtration, washed with water, dried, and recrystallized from ethanol. The percent yields of synthesized compounds **3a–f** are given in Table 7.1.

TABLE 7.1 Synthesized 2-Aminobenzothiazole **3a–f**.

Entry	R_1	R_2	Product name	Yield (%)
3a	–Cl	–H	2-Amino-5-chlorobenzothiazole	80
3b	–H	–OC$_2$H$_5$	2-Amino-6-ethoxybenzothiazole	78
3c	–H	–NO$_2$	2-Amino-6-nitrobenzothiazole	76
3d	–CF$_3$	–H	2-Amino-5-trifluoromethylbenzothiazole	78
3e	–H	–OCH$_3$	2-Amino-6-methoxybenzothiazole	75
3f	–H	–Br	2-Amino-6-bromobenzothiazole	79

7.2.4 SYNTHESIS OF 2-AMINOBENZENETHIOLS (4A–F)

A mixture of appropriate thiazole (0.15 mmol) **3a–f**, potassium hydroxide (2.3 mmol), and water (250 mL) was refluxed until the completion of reaction as confirmed by neutralization of a small amount of the reaction content. After completion of the reaction, refluxing was ceased and reaction contents diluted with water and filtered. The filtrate was cooled below 5 C and neutralized with cold 5N acetic acid. The precipitate obtained was collected by filtration, washed with water, dried, and recrystallized with ethanol or chloroform. The product was protected from air and moisture.

7.2.5 SYNTHESIS OF ZINC MERCAPTIDE OF 2-AMINOBENZENETHIOLES (5A–F)

Appropriate benzenethiol of **3a–f** (0.35 mmol) was dissolved in minimum amount of ethanol. Aqueous sodium hydroxide (0.35 mmol in 5 mL water) was added to the above solution with constant stirring followed by dilution with 50 mL of water. The reaction content was warmed on water bath and filtered. The filtrate was added to a solution of zinc chloride (0.19 mmol) in glacial acetic acid (8 mL) and water (40 mL). Gray-colored suspension appeared immediately. This suspension was boiled for 10 min to facilitate filtration. The product obtained was collected by filtration, washed with water, and dried.

7.2.6 SYNTHESIS OF SUBSTITUTED 3-CHLORO-1,4-NAPHTHOQUINONE 8A–E

A mixture of 2,3-dichloro-1,4-naphthoquinone **6** (0.01 mmol), substituted aniline **7a–e** (0.016 mmol), 2-picoline (0.02 mmol), and ethanol (10 mL) was refluxed for 4 h. The reaction content was cooled and a cherry brown–colored crystalline solid was obtained by filtration, washing with methanol, and drying. The percent yields of synthesized compounds **8a–e** are given in Table 7.2.

TABLE 7.2 Synthesized 1,4-naphthaquinone **8a–e**.

Entry	R_3	R_4	R_5	Product name	Yields (%)
8a	–CH$_3$	–H	–Br	2-(4'-Bromo-2'-methylanilino)-3-chloro-1,4-napthaquinone	85
8b	–CH$_3$	–Cl	–H	2-(3'-Chloro-2'-methylanilino)-3-chloro-1,4-napthaquinone	80
8c	–H	–H	–OCH$_3$	2-(4'-Methoxyanilino)-3-chloro-1,4-napthaquinone	78
8d	–NO$_2$	–H	–OCH$_3$	2-(4'-Methoxy-2'-nitrolanilino)-3-chloro-1,4-napthaquinone	76
8e	–C$_2$H$_5$	–H	–H	2-(2'-Ethylanilino)-3-chloro-1,4-napthaquinone	77

7.2.7 SYNTHESIS OF SUBSTITUTED 6-(4-PHENOXYANILINO)-5H-BENZO[A]PHENOTHIAZINE (9A–N)

A mixture of zinc mercaptide of substituted 2-aminobenzenethiol **5a–f** (0.005 mmol) and substituted 3-chloro-1,4-naphtho-quinone **8a–e** (0.01 mmol) in ionic liquid (5 mL) was stirred for 3 h in a round-bottom flask at 80 °C. The progress of reaction was monitored by TLC Merck silica Gel ^{60}F$_{254}$ sheets and using *n*-hexane:ethyl acetate (8:2) as eluent. After completion of the reaction, the reaction mixture was extracted with ethyl acetate. The combined organic layer was dried over anhydrous Na$_2$SO$_4$ and the solvent was distilled out under reduced pressure. The product so obtained was purified by recrystallization with ethanol. The melting points of all the synthesized compounds **9a–n** are given in Table 7.3.

TABLE 7.3   Melting Points of Compounds **9a–n.**

Entry	Product name	MP (°C)
9a	6-(4-Bromo-2-methylanilino)-10-chloro-*5H*-benzo[*a*] phenothiazin-5-one	269
9b	6-(4-Bromo-2-methylanilino)-9-nitro-*5H*-benzo[*a*]phenothiazin-5-one	281
9c	6-(4-Bromo-2-methylanilino)-9-ethoxy-*5H*-benzo[*a*] phenothiazin-5-one	294
9d	10-Chloro-6-(3-chloro-2-methylanilino)-*5H*-benzo[*a*] phenothiazin-5-one	285
9e	6-(3-Chloro-2-methylanilino)-9-nitro-*5H*-benzo[*a*]phenothiazin-5-one	231
9f	6-(3-Chloro-2-methylanilino)-9-ethoxy-*5H*-benzo[*a*] phenothiazin-5-one	282
9g	6-(4-Methoxyanilino)-10-trifluoromethyl-*5H*-benzo[*a*] phenothiazin-5-one	246
9h	10-Chloro-6-(4-methoxyanilino)-*5H*-benzo[*a*]phenothiazin-5-one	205
9i	6-(4-Methoxyanilino)-9-nitro-*5H*-benzo[*a*]phenothiazin-5-one	214
9j	9-Ethoxy 6-(4-methoxyanilino)-*5H*-benzo[*a*]phenothiazin-5-one	207
9k	9-Methoxy-6-(4-methoxyanilino)-5*H*-benzo[*a*]phenothiazin-5-one	209
9l	9-Bromo-6-(4-methoxyanilino)-5*H*-benzo[*a*]pheno-thiazin-5-one	212
9m	6-(2-Nitroanilino)-9-nitro-*5H*-benzo[*a*]phenothiazin-5-one	280
9n	10-Chloro-6-(2-ethylanilino)-*5H*-benzo[*a*]phenothiazin-5-one	290

7.3   RESULTS AND DISCUSSION

In order to exploit the potential of ionic liquids for the synthesis of benzo[*a*] phenothiazine derivatives, we first carried out the one-pot synthesis of three different ionic liquids, [Bmim]Br, [Bmim]BF$_4$, [Bmim]PF$_6$ using 1-methylimidazole, and various alkyl halides as the starting materials.[132]

The above-synthesized ionic liquids were then employed to achieve our target of synthesis of 5*H*-benzophenothiazines **(9a–n)**. We first prepared zinc mercaptides of 2-aminobenzenethiols **(5a–e)**, which were in turn synthesized from the appropriate 2-aminobenzothiazoles **(4a–e)** using Scheme 7.2.

Substituted 3-chloro-1,4-naphthoquinones **(8a–e)** were then synthesized from 2,3-dichloro-1,4-naphthoquinone **6** and substituted aniline **(7a–e)** (Scheme 7.3).

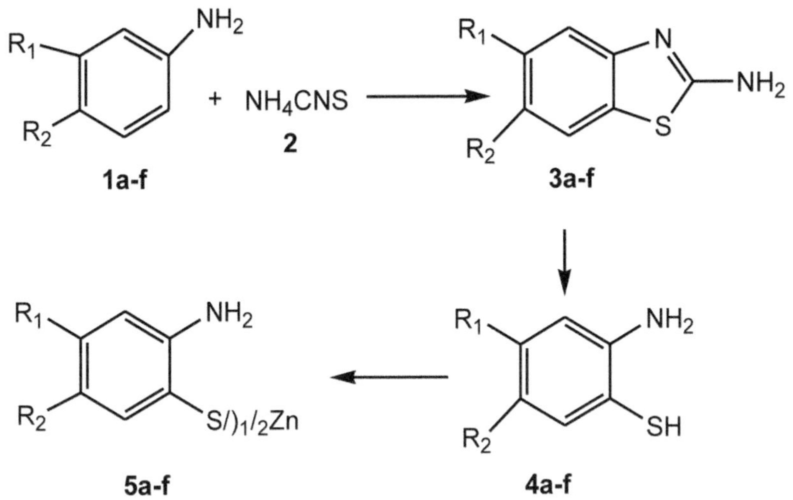

SCHEME 7.2   Synthesis of zinc mercaptide of 2-aminobenzenethiole.

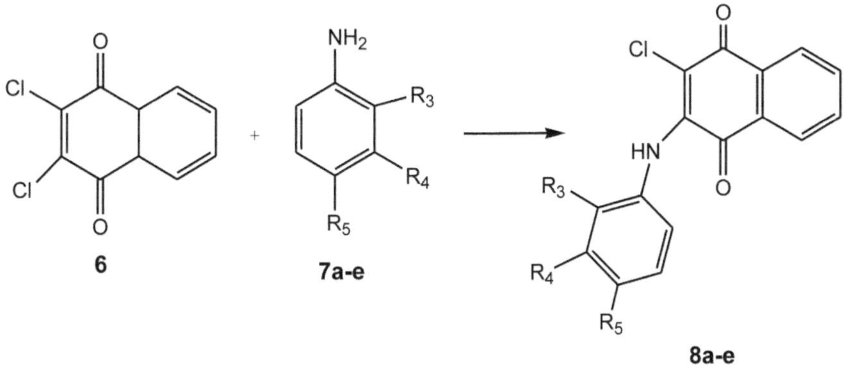

SCHEME 7.3   Synthesis of 1,4-napthaquinones.

To achieve suitable conditions for the synthesis of 5*H*-benzophenothiazines **9(a–n)**, various reaction conditions and catalysts as shown in Table 7.4 were investigated using the reaction of zinc mercaptide of 2-amino-5-chlorobenzothiazole and 2-(4′-bromo-2′-methylanilino)-3-chloro-1,4-napthaquinone as the model reaction (Scheme 7.4).

The above-synthesized zinc mercaptides of 2-aminobenzenethioles **(5a–f)** and 1,4-napthaquinones **(8a–e)** were then reacted together to obtain the final product, that is, 5*H*-benzophenothiazines **(9a–n)** (Scheme 7.5).

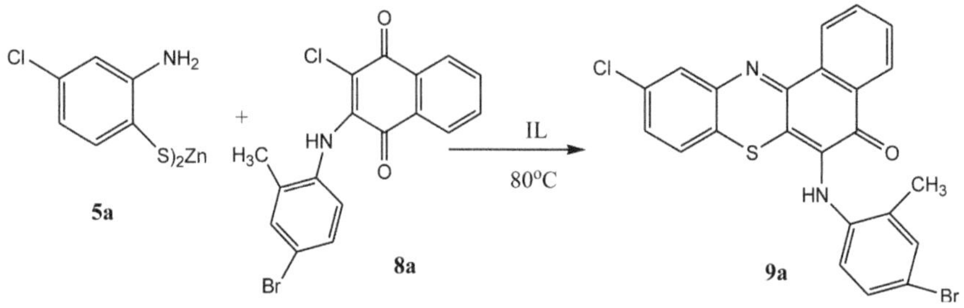

SCHEME 7.4   Model reaction.

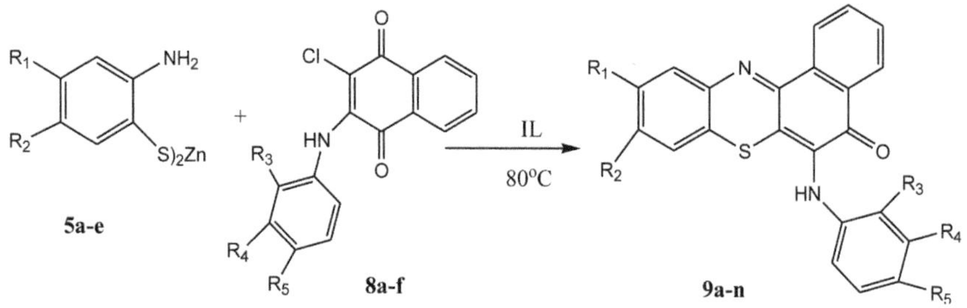

SCHEME 7.5   Synthesis of 5*H*-benzo[*a*]phenothiazine-5-ones.

TABLE 7.4   Effect of Solvents on Synthesis of **9a** Yield (%).

Entry	Solvent	Temp (°C)	Time (h)	Yield (%)
1	Toluene	Reflux	26	17
2	DCM	Reflux	12	20
3	CHCl$_3$	Reflux	6	24
4	Ethyl acetate	Reflux	4.5	29
5	Ethanol	Reflux	5	32
6	Water	Reflux	3.5	39
7	Acetic acid	Reflux	3	42
8	[Bmim]Br	Reflux	2.5	67
9	[Bmim]BF$_4$	Rt	40 min	78
10	[Bmim]PF$_6$	50	5	45
11	[Bmim]PF$_6$	60	4.5	56
12	[Bmim]PF$_6$	70	1.5	79
13	[Bmim]PF$_6$	80	20 min	86

It turned out that the reactions of **5(a–f)** and **8(a–e)** proceeded smoothly in an IL [Bmim][PF$_6$] and gave the corresponding 5*H*-benzophenothiazine **9(a–n)**. The product yield increased remarkably with the temperature until 80 °C beyond which it started decreasing. Interestingly out of the three ILs studied, [Bmim]Br, [Bmim][PF$_6$], and [Bmim][BF$_4$], [Bmim][PF$_6$] gave better results, presumably due to its hydrophobic activation activity. It is postulated that water formed "in situ" from the condensation process is miscible with hydrophilic [Bmim][BF$_4$] and thus detained, which prevents completion of the reaction. In contrast, the hydrophobic nature of [Bmim][PF$_6$] would create a microenvironment to drive the equilibrium by extruding water out of the IL phase and thus result in a higher conversion.

The same reaction was also run in several conventional organic solvents and the results are also included in Table 7.4. Comparing with Toluene, DCM, CHCl$_3$, EtOH, ethyl acetate, and water, ILs exhibited enhanced reactivity by reducing reaction time and improving the yields significantly.

Excellent preliminary results promoted us to further explore the use of ionic liquids in the synthesis of various benzophenothiazines by reaction of different benzenethiols and substituted napthaquinones (Table 7.5). Compounds **9a–n** are stable solids, the structures of which were established by IR, ^{1}H, ^{13}C NMR spectroscopy, and elemental analysis.

TABLE 7.5   Effect of Substituent on Product **(9a–n)** Yield (%) in [BMIM]PF$_6$.

Product	R$_1$	R$_2$	R$_3$	R$_4$	R$_5$	Yield (%)	Time (h)	Refs.
9a	Cl	H	CH$_3$	H	Br	69	5	–
9b	H	OC$_2$H$_5$	CH$_3$	H	Br	78	5	–
9c	H	NO$_2$	CH$_3$	H	Br	72	5	–
9d	Cl	H	CH$_3$	Cl	H	86	4	–
9e	H	OC$_2$H$_5$	CH$_3$	Cl	H	83	4	–
9f	H	NO$_2$	CH$_3$	Cl	H	81	4	–
9g	CF$_3$	H	H	H	OCH$_3$	80	3.5	[128]
9h	Cl	H	H	H	OCH$_3$	82	4	[128]
9i	H	NO$_2$	H	H	OCH$_3$	78	3	[128]
9j	H	OC$_2$H$_5$	H	H	OCH$_3$	76	4	[128]
9k	H	OCH$_3$	H	H	OCH$_3$	80	5	[128]
9l	H	Br	H	H	OCH$_3$	85	4	[128]
9m	H	NO$_2$	NO$_2$	H	OCH$_3$	75	5	[129]
9n	Cl	H	C$_2$H$_5$	H	H	82	4	[129]

A plausible mechanism for this condensation process is proposed in Scheme 7.6. IL enhances the rate of this condensation reaction probably due to its Bronsted acidity, which can be attributed to the hydrogen atom of imidazolium cation that interacts with the heteroatoms, thereby increasing polarization. The intermediate **III** is formed by an initial nucleophilic attack by thiol ion **I** on the three-position of **II** thus losing one of the halogen atoms and diaryl sulfide **IV** is formed. Cyclization then occurs via internal condensation of the amino group with the carbonyl in diaryl sulfide resulting in the release of a water molecule and the formation of *5H*-benzophenothiazene derivatives **V**.

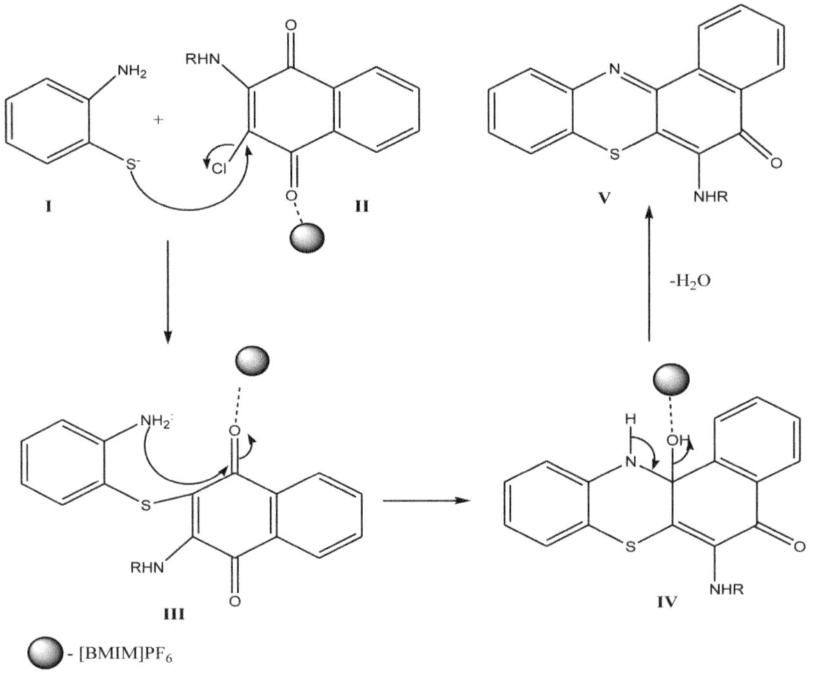

SCHEME 7.6   Plausible mechanism.

Recovery and reuse of [Bmim][PF$_6$] was also studied. Upon completion of the condensation process, product was obtained through extraction with diethyl ether and the remaining IL phase was recycled in subsequent reactions. Further studies showed that the recovered [Bmim][PF$_6$] could be successfully recycled for at least 5 times without obvious loss in its efficiency (Table 7.6).

TABLE 7.6 Studies on Recovery and Reuse of [Bmim][PF$_6$].

Cycles	Time (min)	Temp (°C)	Yield (%)
1.	20	75	86
2.	20	80	85
3.	20	80	85
4.	20	80	84
5.	20	80	82

All the compounds **9a–n** are stable solids, the structures of which were established by IR, ^{1}H, ^{13}C NMR spectroscopy, and elemental analysis.

7.3.1 SPECTRAL STUDIES

The IR spectra of compound **9a–n** showed a strong band in the region 1656–1640 cm^{-1} due to >C=O group and >NH stretching vibration appeared in the region 3225–3185 cm^{-1} as a weak band. The shift of these two stretching vibrations toward lower frequency may be attributed to two well-established facts viz. intramolecular hydrogen bonding through >N–H...O=C< and ionic resonance effect as shown next.

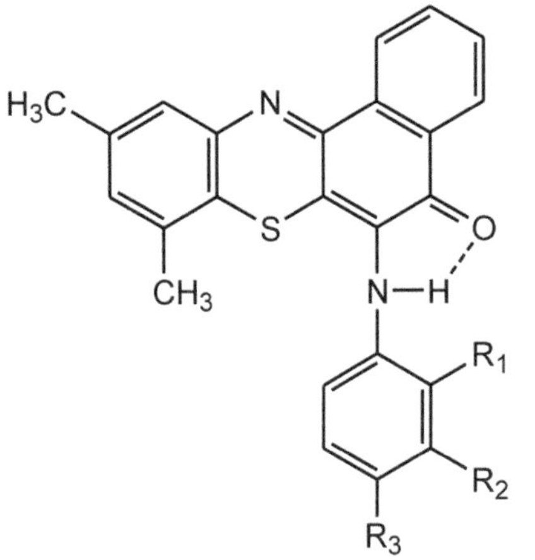

Intramoleculer hydrogen bonding

The structure of compounds **9a–n** reveals that sulfur atom is present at the β-position with respect to the carbonyl group and due to electron-donating tendency of sulfur atom, the resonance effect was observed. Therefore, the bond length of carbonyl group increased and thus decreases the force constant, which results in the lowering of wave number of $>C=O$ group.

The appearance of weak absorption band in the range 665–636 cm^{-1} was due to C–S–C linkage. Bands between 1527–1514 cm^{-1} and 1365–1343 cm^{-1} were assigned to C=N and C–N stretching vibrations, respectively. The C–H aliphatic and aromatic stretching vibrations were observed at 2895–2855 cm^{-1} and 3110–3035 cm^{-1}, respectively.

In the ^{1}H NMR spectra of synthesized compounds **9a–n**, the multiplet for aromatic protons appeared in the region δ 6.57–8.28. The >NH proton signal of anilino group in compounds **9a–n** was found to be merged with the aromatic proton signals. The protons of ethoxy, ethyl, methoxy, and methyl groups also showed their peaks at their respective positions.

^{13}C-NMR data of compounds **9a–n** showed peaks for aromatic carbons in the range 112–158 ppm, in addition to the ethoxy, ethyl, methoxy, and methyl signals at their respective positions.

7.3.2 CHARACTERIZATION DATA OF SYNTHESIZED COMPOUNDS (9A–N)

7.3.2.1 6-(4-BROMO-2-METHYL-PHENYLANILINO)-10-CHLORO-5H-BENZO[A]PHENO-THIAZIN-5-ONE, 9A

IR (KBr) (v_{max}/cm^{-1}), 3040 (aromatic hydrogen), 3225 (NH stretching), 2872 (aliphatic hydrogen), 1650 (C=O), 1520 (C=N), 1350 (C–N), 765 (C–Cl),

660 (C–S–C); ^{1}H NMR (CDCl$_3$): δ 6.74–8.19 (m, 10H, Ar–H, and NH anilino), 2.30 (s, 3H, CH$_3$), ^{13}C NMR (CDCl$_3$): δ 187.05, 163.52, 153.30, 146.42, 136.15, 133.92, 133.56, 132.27, 131.32, 130.28, 129.71, 128.94, 128.87, 128.43, 128.22, 127.07, 127.28, 126.43, 126.08 120.05, 117.13, 113.59, 11.52. Anal. calcd. for C$_{23}$H$_{14}$BrClN$_2$OS, C, 57.34; H, 2.93; N, 5.81. Found C, 57.21; H, 2.46; N, 5.76.

7.3.2.2 6-(4-BROMO-2-METHYL-PHENYLANILINO)-9-NITRO-5H-BENZO[A]PHENO-THIAZIN-5-ONE, 9B

IR (KBr): (v$_{max}$/cm^{-1}) 3100 (aromatic hydrogen), 3220 (NH stretching), 2870 (aliphatic hydrogen), 1645 (C=O), 1520 (C=N), 1480 (NO$_2$), 1360 (C–N), 760 (C–Br), 632 (C–S–C); ^{1}H NMR (CDCl$_3$): δ 6.89–8.19 (m, 10H, Ar–H, and NH anilino), 2.30 (s, 3H, CH$_3$), ^{13}C NMR (CDCl$_3$): δ 187.05, 163.52, 153.64, 146.41, 146.35, 136.11, 133.90, 133.55, 132.23, 132.17, 131.41, 131.34, 130.27, 129.74, 128.82, 128.45, 126.42, 122.43, 120.07, 117.65, 117.10, 113.57, 11.59, Anal. calcd. for C$_{23}$H$_{14}$BrN$_3$O$_3$S, C, 56.11; H, 2.87; N, 8.53. Found C, 56.04; H, 2.43; N, 8.25.

7.3.2.3 6-(4-BROMO-2-METHYL-PHENYLANILINO)-9-ETHOXY-5H-BENZO[A]PHENO-THIAZIN-5-ONE, 9C

IR (KBr) (v$_{max}$/cm^{-1}), 3110 (aromatic hydrogen), 3215 (NH stretching), 2855 (aliphatic hydrogen), 1644 (C=O), 1525 (C=N), 1365 (C–N), 1243, 760 (C–Br), 661 (C–S–C); ^{1}H NMR (CDCl$_3$): δ 6.90–8.35 (m, 10H, Ar–H, and NH anilino), 3.50 (q, 2H, OCH$_2$), 1.26 (t, 3H, CH$_3$), 2.30 (s, 3H, CH$_3$), ^{13}C NMR (CDCl$_3$): δ 187.01, 163.63, 155.94, 153.42, 145.27, 132.19, 130.10, 120.01, 116.45, 112.23, 136.21, 133.03, 132.19, 131.32, 130.24, 128.75, 128.91, 128.45, 125.36, 116.22, 112.08, 107.04, 64.91, 13.33, 11.50, Anal. calcd. for C$_{25}$H$_{19}$BrN$_2$O$_2$S, C, 61.10; H, 3.90; N, 5.70. Found C, 61.98; H, 3.88; N, 5.43.

7.3.2.4 10-CHLORO-6-(3-CHLORO-2-METHYLPHENYLANILINO)-5H-BENZO[A]PHENO-THIAZIN-5-ONE, 9D

IR (KBr) (v$_{max}$/cm^{-1}), 3038 (aliphatic hydrogen), 3221 (NH stretching), 2890 (aromatic hydrogen), 1645 (C=O), 1520 (C=N), 1355 (C–N), 760

(C–Cl), 647 (C–S–C); ^{1}H NMR (CDCl$_3$): δ 6.70–8.20 (m, 10H, Ar–H, and NH aniline), 2.30 (s, 3H, CH$_3$), ^{13}C NMR (CDCl$_3$): δ 187.04, 163.51, 153.38, 148.86, 136.13, 135.51, 133.50, 132.22, 131.37, 130.23, 128.90, 128.87, 128.44, 128.27, 127.82, 127.66, 127.14, 126.71, 124.76, 120.04, 118.73, 113.56, 3.10. Anal. calcd. for C$_{23}$H$_{14}$Cl$_2$N$_2$OS, C, 63.16; H, 3.23; N, 6.41. Found C, 63.03; H, 3.11; N, 6.17.

7.3.2.5 6-(3-CHLORO-2-METHYLPHENYLANILINO)-9-NITRO-5H-BENZO[A]PHENO-THIAZIN-5-ONE, 9E

IR (KBr) (v$_{max}$/cm^{-1}), 3105 (aliphatic hydrogen), 3205 (NH stretching), 2895 (aromatic hydrogen), 1642 (C=O), 1525, 1467 (NO$_2$), 1365 (C–N), 760 (C–Cl), 653 (C–S–C); ^{1}H NMR (CDCl$_3$): δ 6.42–8.259 (m, 10 H, Ar–H, and NH anilino), 2.30 (s, 3H, CH$_3$), ^{13}C NMR (CDCl$_3$): δ 187.03, 163.59, 154.82, 148.80, 146.53, 136.12, 135.54, 133.05, 132.26, 132.11, 131.45, 131.37, 130.23, 128.86, 128.42, 127.68, 124.75, 122.33, 120.07, 118.72, 117.39, 113.56, 3.10. Anal. calcd. for C$_{23}$H$_{14}$ClN$_3$O$_3$S, C, 61.68; H, 3.51; N, 9.38, Found C, 61.22; H, 3.11; N, 9.32.

7.3.2.6 6-(3-CHLORO-2-METHYLPHENYLANILINO)-9-ETHOXY-5H-BENZO[A]PHENOTHIAZIN-5-ONE, 9F

IR (KBr) (v$_{max}$/cm^{-1}), 3112 (aromatic hydrogen), 3220 (NH stretching), 2896 (aliphatic hydrogen) 1645 (C=O), 1523 (C=N), 1360 (C–N), 1235 (C–O–C), 765 (C–Cl), 657 (C–S–C); ^{1}H NMR (CDCl$_3$): δ 6.35–8.20 (m, 10H, Ar–H, and NH anilino) 3.50 (q, 2H, OCH$_2$), 1.26 (t, 3H, OCH$_2$CH$_3$), 2.30 (s, 3H, CH$_3$), ^{13}C NMR (CDCl$_3$): δ 187.05, 163.52, 156.93, 154.32, 148.86, 136.15, 135.53, 133.59, 132.02, 131.45, 131.34, 130.28, 128.18, 128.43, 127.64, 124.73, 120.03, 118.72, 117.65, 113.02, 113.57, 108.02, 65.11, 14.35, 3.13. Anal. calcd. for C$_{25}$H$_{19}$ClN$_2$O$_2$S, C, 67.18; H, 4.28; N, 6.27. Found C, 67.09; H, 4.22; N, 6.24.

7.3.2.7 6-(4-METHOXYANILINO)-10-TRIFLUROMETHYL-5H-BENZO[A]PHENOTHIAZIN-5-ONE, 9G

IR (KBr): (v$_{max}$/cm^{-1}) 3100 (aromatic hydrogen), 3220 (NH stretching), 2870 (aliphatic hydrogen), 1645 (C=O), 1520 (C=N), 1360 (C–N), 760

(C–Br), 632 (C–S–C); ^{1}H NMR (CDCl$_3$):δ 6.45–8.98 (m, 11H, Ar–H, and –NH anilino), 3.40 (s, 3H, –OCH$_3$); ^{13}C NMR (CDCl$_3$): δ 187, 164.6, 154.2, 152, 137.2, 134.4, 132.4, 131.3, 130.8, 129.8, 129.5, 129.3, 129.1, 124.1, 119.3, 116.1, 114.9, 56. Anal. calcd for C$_{24}$H$_{15}$F$_3$N$_2$O$_2$S: C, 63.71; H, 3.34; N, 6.19; S, 7.09. Found: C, 63.62; H, 3.29; N, 6.08; S, 6.99%.

7.3.2.8   10-CHLORO-6-(4-METHOXYANILINO)-5H-BENZO[A] PHENOTHIAZIN-5-ONE, 9H

IR (KBr): (v_{max}/cm^{-1}) 3120 (aromatic hydrogen), 3250 (NH stretching), 2860 (aliphatic hydrogen), 1648 (C=O), 1525 (C=N), 1362 (C–N), 762 (C–Br), 625 (C–S–C); ^{1}H NMR (CDCl$_3$): δ 6.30–8.29 (m, 11H, Ar–H, and –NH anilino), 3.34 (s, 3H, –OCH$_3$); ^{13}C NMR (CDCl$_3$): δ 188, 164.8, 157.2, 152, 141, 134.7, 132.6, 133.7, 132.4, 129.1, 129.8, 124.3, 123.2, 116.6, 114.2, 57. Anal. calcd for C$_{23}$H$_{15}$ClN$_2$O$_2$S: C, 65.95; H, 3.61; N, 6.69; S, 7.65. Found: C, 65.97; H, 3.33; N, 6.59; S, 7.81%.

7.3.2.9   6-(4-METHOXYANILINO)-9-NITRO-5H-BENZO[A] PHENOTHIAZIN-5-ONE, 9I

IR (KBr): (v_{max}/cm^{-1}) 3090 (aromatic hydrogen), 3270 (NH stretching), 2866 (aliphatic hydrogen), 1655 (C=O), 1528 (C=N), 1470 (NO$_2$), 1350 (C–N), 766 (C–Br), 630 (C–S–C); ^{1}H NMR (CDCl$_3$): δ 6.21–8.46 (m, 11H, Ar–H, and –NH anilino), 3.47 (s, 3H, –OCH$_3$); ^{13}C NMR (CDCl$_3$): δ 190, 164.4, 157.8, 155, 146.9, 139, 137.7, 135.4, 132.7, 131.0, 131.8, 129.4, 129, 122.3, 117.7, 115, 114.2, 58. Anal. calcd for C$_{23}$H$_{15}$N$_3$O$_4$S: C, 64.33; H, 3.52; N, 9.78; S, 7.47. Found: C, 64.27; H, 3.59; N, 9.42; S, 7.89%.

7.3.2.10   9-ETHOXY-6-(4-METHOXYANILINO)-5H-BENZO[A] PHENOTHIAZIN-5-ONE, 9J

IR (KBr): (v_{max}/cm^{-1}) 3170 (aromatic hydrogen), 3214 (NH stretching), 2876 (aliphatic hydrogen), 1638 (C=O), 1517 (C=N), 1362 (C–N), 756 (C–Br), 630 (C–S–C); ^{1}H NMR (CDCl$_3$): δ 6.23–8.71 (m, 11H, Ar–H, and –NH anilino), 3.45 (s, 3H, –OCH$_3$); 1.22 (t, 3H, –CH$_3$), 3.56 (q, 2H, –CH$_2$–CH$_3$); ^{13}C NMR (CDCl$_3$): δ 188, 164.2, 155.2, 153.9, 153, 138.6,

137.7, 135.2, 133.8, 132.5, 132.6, 130.6, 130.1, 118.3, 116.7, 115.2, 115, 111.2, 66.7, 54, 15.2. Anal. calcd for $C_{25}H_{20}N_2O_3S$: C, 70.07; H, 4.70; N, 6.54; S, 7.48. Found: C, 70.11; H, 4.53; N, 6.57; S, 7.48%.

7.3.2.11   9-METHOXY-6-(4-METHOXYANILINO)-5H-BENZO[A] PHENOTHIAZIN-5-ONE, 9K

IR (KBr): (v_{max}/cm^{-1}) 3130 (aromatic hydrogen), 3215 (NH stretching), 2865 (aliphatic hydrogen), 1652 (C=O), 1517 (C=N), 1353 (C–N), 768 (C–Br), 639 (C–S–C); ^{1}H NMR (CDCl$_3$): δ 6.41–8.48 (m, 11H, Ar–H, and –NH anilino), 3.42 (s, 3H, –OCH$_3$); 3.49 (s, 3H, –OCH$_3$); ^{13}C NMR (CDCl$_3$): δ 188.8, 166.4, 160.2, 155.8, 154, 139.2, 137, 135.6, 132.8, 131.2, 130.7, 127.8, 125.4, 121.6, 116.7, 115, 113.8, 55. Anal. calcd for $C_{24}H_{18}N_2O_3S$: C, 69.55; H, 4.38; N, 6.76; S, 7.74. Found: C, 69.63; H, 4.19; N, 6.64; S, 7.58%.

7.3.2.12   9-BROMO-6-(4-METHOXYANILINO)-5H-BENZO[A] PHENOTHIAZIN-5-ONE, 9L

IR (KBr): (v_{max}/cm^{-1}) 3160 (aromatic hydrogen), 3320 (NH stretching), 2868 (aliphatic hydrogen), 1635 (C=O), 1527 (C=N), 1340 (C–N), 776 (C–Br), 640 (C–S–C); ^{1}H NMR (CDCl$_3$): δ 6.39–8.47 (m, 11H, Ar–H, and –NH anilino), 3.47 (s, 3H, –OCH$_3$); ^{13}C NMR (CDCl$_3$): δ 189, 165.1, 157.3, 155, 138, 137.6, 136.7, 132, 131.5, 130.7, 129, 129.1, 128.2, 125.4, 120.7, 116.7, 114.4, 56.8. Anal. calcd for $C_{23}H_{15}BrN_2O_2S$: C, 59.62; H, 3.26; N, 6.05; S, 6.92. Found: C, 59.68; H, 3.15; N, 6.13; S, 6.75%.

7.3.2.13   6-(4-METHOXY-2-NITROPHENYLANILINO)-9-NITRO-5H-BNZO[A]PHENOTHIAZIN-5-INE, 9M

IR (KBr) (v_{max}/cm^{-1}), 3110 (aromatic hydrogen), 3212 (NH stretching), 2895 (aromatic hydrogen) 1640 (C=O), 1530, 1465 (NO$_2$), 1362 (C–N), 1254 (C–O–C), 651 (C–S–C); ^{1}H NMR (CDCl$_3$): δ 6.38–8.40 (m, 10H, Ar–H, and NH anilino), 3.41 (s, 3H, OCH$_3$), ^{13}C NMR (CDCl$_3$): δ 187.01, 163.54, 153.34, 152.93, 146.50, 136.12, 136.07, 134.19, 133.52, 132.29, 132.15, 131.56, 131.30, 130.42, 128.83, 128.45, 122.54, 121.38, 120.02,

117.81, 117.4, 110.09, 56.03. Anal. calcd. for $C_{23}H_{14}N_4O_6S$, C, 58.22; H, 2.97; N, 11.81 Found C, 58.15; H, 2.89; N, 11.72.

7.3.2.14 6-(2-ETHYL-PHENYLANILINO)-9-NITRO-5H-BENZO[A] PHENOTHIAZIN-5-ONE, 9N

IR (KBr) (v_{max}/cm^{-1}), 3104 (aromatic hydrogen), 3205 (NH stretching), 2870 (aliphatic hydrogen), 1642 (C=O), 1523, 1475 (NO$_2$), 1365 (C–N), 642 (C–S–C); ^{1}H NMR (CDCl$_3$): δ 6.74–8.21 (m, 11H, Ar–H, and NH anilino), 2.51 (q, 2H, CH$_2$), 1.21 (t, 3H, CH$_3$), ^{13}C NMR (CDCl$_3$): δ 187.01, 163.63, 153.15, 145.51, 144.53, 136.82, 133.42, 131.76, 132.35, 132.24, 131.14, 130.21, 128.88, 128.77, 128.47, 126.96, 126.24, 122.49, 120.02, 118.43, 117.45, 115.01, 19.61, 16.18, Anal. calcd. for C_{24}, $H_{17}N_3O_3S$, C, 67.43; H, 4.01; N, 9.83. Found C, 67.27; H, 3.99; N, 9.80.

7.4 CONCLUSIONS

In conclusion, we have demonstrated a highly versatile and efficient synthetic methodology for the transformation of zinc mercaptides of 2-aminobenzenethiols and substituted 1,4-napthaquinones into 5*H*-benzo[*a*] phenothiazine-5-ones. Most reactions proceeded to completion in 4–5 h with very simple workup procedure and direct isolation of the products in good yields. The characterization of all the synthesized compounds was done by IR, ^{1}HNMR, and ^{13}CNMR spectroscopy.

CONFLICTS OF INTEREST

The authors confirm that the article has no conflict of interest.

ACKNOWLEDGMENTS

The authors are thankful to CSIR, New Delhi for the award of fellowships (SRF) to the authors. The authors are also grateful to MNIT, Jaipur for the spectral analysis of the compounds synthesized.

KEYWORDS

- heterocyclic compounds
- ionic liquids
- phenothiazine
- benzothiazole
- napthaquinone

REFERENCES

1. Arora, P.; Arora, V.; Lambha, H. S.; Wadhwa, D. *Int. J. Pharm. Sci. Res.* **2012**, *3*, 2947.
2. Eicher, T.; Hauptmann, S. *The Chemistry of Heterocycles: Structure, Reactions, Synthesis and Applications*; 2nd ed.; Wiley-VCH Verlag GmbH: Weinheim, 2003.
3. Maruthamuthu; Rajam, S.; Christina, R. S. P.; Bharathi, A. G.; Ranjith, R. *J. Chem. Pharm. Res.* **2016**, *8* (5), 505.
4. IUPAC Compendium of Chemical Terminology—The Gold Book Heterocyclic Compounds, 2009. http://goldbook.iupac.org/H02798.html.
5. Quin, L. D.; Tyrell, J. A. *Fundamentals of Heterocyclic Chemistry*; John Wiley & Sons, Inc.: Hoboken, NJ, 2010.
6. Martins, P.; Jesus, J.; Santos, S.; Raposo, L. R.; Roma-Rodrigues, C.; Baptista, P. V.; Fernandes, A. R. *Molecules* **2015**, *20*, 16852.
7. Balaban, A. T. *Chem. Rev.* **2004**, *104*, 2777.
8. Czarnik, A. *Acc. Chem. Res.* **1996**, *29*, 112.
9. Hasko, G.; Linden, J.; Pacher, P. *Nat. Rev. Drug Discovery* **2008**, *7*, 759.
10. Muhammad, Z. A.; Edrees, M. M.; Faty, R. A. M.; Gomha, S. M.; Alterary, S. S.; Mabkhot, Y. N. *Molecules* **2017**, *22*, 1211.
11. Al-Mulla, A. *Der Pharma Chemica* **2017**, *9*, 141.
12. Andrade, C. K. Z.; Alves, L. M. *Curr. Org. Chem.* **2005**, *9*, 195.
13. Ranu, B. C.; Saha, A.; Dev, A. *Curr. Opin. Drug Discovery Dev.* **2010**, *13*, 658.
14. Lakner, F. J.; Negrete, G. R. *Synlett* **2002**, *4*, 0643.
15. Lo, A. S. W.; Horváth, I. T. *Green Chem.* **2015**, *17*, 4701. Top of for Bottom of Form
16. Welton, T. *Chem. Rev.* **1999**, *99*, 2071.
17. Meine, N.; Benedito, R. *Green Chem.* **2010**, *12*, 1711.
18. Fuller, J.; Carlin, R. T.; Osteryoung, R. A. *J. Electrochem. Soc.* **1997**, *144*, 3881.
19. Rantwijik, F. V.; Sheddon, R. A. *Chem. Rev.* **2007**, *107*, 2757.
20. Singh, G.; Kumar, A. *Ind. J. Chem.* **2008**, *47* (A), 495.
21. Earle, M. J.; Seddon, K. R. *Pure Appl. Chem.* **2000**, *72*, 1391.
22. Marcus, Y. *Ionic Liquid Properties-from Molten Salts to RTILs*; Springer International Publishing: Switzerland, 2016.
23. Ahrens, S.; Peritz, A.; Strassner, T. *Angew. Chem. Int. Ed.* **2009**, *48*, 7908.

24. Fremantle, M. *Chem. Eng. News* **1998**, *76*, 32.
25. Chiappe, C.; Malvaldi, M.; Pomelli, C. S. *Pure Appl. Chem.* **2009**, *81*, 767.
26. Pandey, S. *Anal. Chim. Acta* **2006**, *556*, 38.
27. Hajipour, A. R.; Rafiee, F. *Org. Prep. Proced. Int.* **2015**, *47*, 249.
28. Rogers, R. D.; Seddon, K. R. *Science* **2003**, *302*, 792.
29. Earle, M. J.; Seddon, K. R. *Pure Appl. Chem.* **2000**, *72*, 1391.
30. Dupont, J.; de Souza, R. F.; Suarez, P. A. Z. *Chem. Rev.* **2002**, *102*, 3667.
31. Migowski, P.; Dupont, J. *Chem. Eur. J.* **2007**, *13*, 32.
32. Dupont, J.; Fonseca, G. S.; Umpierre, A. P.; Fichtner, P. F. P.; Teixeira, S. R. *J. Am. Chem. Soc.* **2002**, *124*, 4228.
33. Massie, S. P. *Chem. Rev.* **1954**, *54*, 791.
34. Okafor, C. O. *Int. J. Sulfur Chem.* **1971**, *6*, 237.
35. Okafor, C. O. *J. Heterocycl. Chem.* **1981**, *18*, 405.
36. Okafor, C. O.; Okoro, U. C. *Dyes Pigm.* **1988**, *9*, 422.
37. Kym, O. *Ber. Dtsch. Chem. Ges.* **1980**, *23*, 2458.
38. Lauth, C. *Chem. Ber.* **1876**, *9*, 1035.
39. Berthensen, A. *Chem. Ber.* **1883**, *16*, 2896.
40. Curry; S. M. *Drug Psychiatry* **1985**, *3*, 79.
41. Maafi, M.; Aaron, J. J.; Mahedero, M. C.; Salinas, F. *Appl. Spectrosc.* **1998**, *52*, 91.
42. King, W. B.; Nanya, S.; Yamaguchi, Y.; Maekawa, E.; Ueno, Y. *J. Heterocycl. Chem.* **1987**, *24*, 91.
43. Kubo, Y.; Kuwana, M.; Yoshiba, K. *Chem. Expr.* **1987**, *3*, 663.
44. Girard, Y.; Hamel, P.; Therien, M.; Springer, J. P.; Hishfield, J. *J. Org. Chem.* **1987**, *52*, 4000.
45. Nauya, S.; Maekawa, E.; Kang, W. B.; Ueno, Y. *J. Heterocycl. Chem.* **1986**, *23*, 589.
46. Galbraith, F.; Smiles, S. *J. Chem. Soc.* **1935**, 1234.
47. Clerq, E. D. *Nucleosides Nucleotides* **1985**, *4*, 3.
48. Gupta, R. R. *Phenothiazines and 1,4-Benzothiazines: Chemical and Biomedical Aspects Chem. Bio. Aspects*; Elsevier: Amsterdam, 1988.
49. Pumima, A.; Mathur, N.; Gupta, V.; Ojha, K. A. *Pharmacia* **1991**, *46*, 885.
50. Yogeshi, D.; Rahul, D.; Naveen, G.; Gowlam, D. C. *Eur. J. Chem.* **2008**, *5*, 1063.
51. Mieczyslaw, L. *Stizody* **2007**, *9*, 44.
52. Mosnaim, A.; Ranade, V. *Am. J. Ther.* **2006**, *13*, 261.
53. Motohashi, N.; Kawase, M.; Saito, S.; Kurihara, T.; Satoh, K.; Nakashima, H.; Premanathan, M.; Arakaki; Sakagami, H.; Molnar, J. *Int. J. Antimicrob. Agents* **2000**, *14*, 203.
54. Bansode, T. N.; Shelke, J. V.; Dongre, V. G. *Eur. J. Med. Chem.* **2009**, *44*, 5094.
55. Ordway, D.; Viveiros, M.; Leandro, C.; Arroz, M. J.; Amaral, L. *Int. J. Antimicrob. Agents* **2002**, *20*, 34.
56. Kubota, K.; Kurebayashi, H.; Miyachi, H.; Tobe, M.; Onishi; Isobe, Y. *Bioorg. Med. Chem. Lett.* **2009**, *19*, 2766.
57. Ondarza, R. N.; Iturbe, A.; Hernandez, E.; Tamayo, E. M.; Hurtado, G. *Arch. Med. Res.* **2000**, *31*, 512.
58. Gindon, Y.; Girard, Y.; Lau, C. K.; Fortin, R.; Rokach, J.; Yoakim, C. (Merck Frosst Canada Inc.) U.S. U54, 611, 056 (Cl. 544-31; CO 7D26 5/38), 09 Sep, (1986), US Appl. 539, 215, 05 Oct. (1983), 25PP; *Chem. Abstr.* **1987**, *106*, 33082.

59. Sakagami, K.; Takahashi, H.; Yoshida, H.; Yamamura, M. *Anticancer Res.* **1995**, *6B*, 2533.
60. Motohashi, N.; Sakagami, H.; Kamata, K.; Yamamoto, Y. *Anticancer Res.* **1991**, *11*, 1933; *Chem. Abstr.* **1992**, *116*, 165862.
61. Tanaka, M.; Wayda, K.; Molnar, J.; Parkanyi, C.; Aaron, J. J.; Motohashi, N. *Anticancer Res.* **1997**, *17*, 839.
62. Tanaka, M.; Wayda, K.; Molnar, J.; Motohahi, N. *Anticancer Res.* **1996**, *16* (6B), 3625.
63. Silva, G. A.; Costa, L. M. M.; Miranda, A. *Bioorg. Med. Chem.* **2004**, *12*, 3149.
64. Singh, G.; Swati, S. A. K.; Prakash, L. *Indian J. Heterocycl. Chem.* **1996**, *6*, 9.
65. M. Frost Canada, Inc., Jpn Kokkai Koho JP 59 144 774 [84 144 774] (Cl. CO7D 265/38) **1984**, US Appl. 459, **1983**, 924; *Chem. Abstr.* **1985**, *102*, 6515.
66. Molnar, J.; Mandi, Y.; Kiraly, J. *Acta Microbiol. Acad. Sci. Hung.* **1976**, *23*, 45.
67. Shukla, S.; Prakash, L. *Phosphorus Sulfur Silicon* **1995**, *102*, 39.
68. Sharma, R.; Goyal, R. D.; Prakash, L. *Phosphorus Sulfur Silicon* **1993**, *80*, 23.
69. Candurra, N. A.; Maskin, L.; Pamonte, E. B. *Antiviral Res.* **1996**, *31*, 149.
70. Mucsi, I.; Molnar, J. *Enlight Assoc.* **1995**, 347.
71. Mucsi, I.; Molnar, J.; Motohashi, N. *Int. J. Antimicrob. Agents* **2001**, *18*, 67.
72. Motohashi, N.; Sakagami, H.; Kurihara, T.; Csuri, C.; Molnar, J. *Anticancer Res.* **1992**, *12*, 135.
73. Lin, A. J.; Guan, J.; Kyle, D. E.; Milhous, W. K. (United States Army Medical Research and Material Command, USA) PCT Int. Appl. WO 02, 89, 810 (Cl. A61 K31/5415), 14, 2002; *Chem. Abstr.* **2002**, *137*, 25.
74. Shirley, D. A.; Sen, K.; Gilmer, J. C. *J. Org. Chem.* **1962**, *26*, 3587.
75. Shirley, D. A.; Gilmer, J. C.; Waters, W. D. *J. Chem. Soc.* **1964**, *IV*, 5260.
76. Wakelin, L. P. G.; Adams, A.; Denny, W. A. *J. Med. Chem.* **2002**, *45*, 894.
77. Satoh, K.; Sakagami, H.; Motohashi, N. *Anticancer Res.* **1997**, *17* (4A), 2539.
78. Badger, G. M. *Proc. R. Soc.* **1942**, *B130*, 255.
79. Motohashi, N. *Yakugaku Zasshi* **1983**, *13*, 317.
80. Foley, J. W.; Cincotta, L. U.S. US 4, 962, 197, (Cl. 544-31; CO7D 265/30), 09 Oct. **1990**, Appl. 157, 214, 12 Feb. 1988, 10PP, *Chem. Abstr.* **1991**, *114*, 185524.
81. Kurihara, T.; Motohashi, N.; Pang, G. L.; Higano, M.; Kiguchi, K.; Molnar, J. *Anticancer Res.* **1996**, *16* (5A), 2757; *Chem. Abstr.* **1997**, *126*, 27894.
82. Hendrzak-Henion, J. A.; Knisely, T. L.; Cincotta, L.; Cinotta, A. H. *Photochem. Photobiol.* **1999**, *69* (5), 575; *Chem. Abstr.* **1999**, *131*, 84975.
83. Motohashi, N.; Kurihara, T.; Satoh, K.; Sakagami, H.; Musci, I.; Pusztai, R.; Szabo, M.; Molnar, J. *Anticancer Res.* **1999**, *19* (3A), 1837; *Chem. Abstr.* **2000**, *132*, 116953.
84. Motohashi, N.; Kurihara, T.; Yamanaka, W.; Satoh, K.; Sakagami, H.; Molnar, J. *Anticancer Res.* **1997**, *17*, 3431.
85. Motohashi, N.; Kawase, M.; Saito, S.; Sakajami, H. *Curr. Drug Targets* **2000**, *1*, 237.
86. Motohashi, N.; Sakagami, H.; Kamata, K.; Yamamoto, Y. *Anticancer Res.* **1991**, *11*, 1933.
87. Noboru, M.; Masami, K.; Setsuo, S.; Teruo, K.; Kazuc, S.; Hideki, N.; Mariappan, P.; Rieko, A.; Hiroshi, S.; Joseph, M. *Int. J. Antimicrob. Agents* **2000**, *14*, 203.
88. Motohashi, N.; Kawase, M.; Satoh, K.; Sakajami, H. *Curr. Drug Target* **2006**, *7*, 1055.
89. Molnar, J.; Mandi, Y.; Petri, I.; Petofi, S.; Sakagami, H.; Kurihara, T.; Motohashi, N. *Anticancer Res.* **1993**, *13* (2), 439.

90. Janneke, W.; Henry, V. V.; Mark, H. P. V.; Silvina, A. F.; Martinl, J. S.; Iwan, J. P. de.; Rob, L. *Biomol. Chem.* **2006**, *14*, 7213.

91. Mosnaim, D. A.; Ronade, V. V.; Wolf, E. M.; Puente, J.; Antonieta, M. *Am. J. Ther.* **2006**, *13*, 261.

92. Bisi, A.; Meli, M.; Gobbi, S.; Rampa, A.; Tolomeo, M.; Dusonchet, L. *Bioorg. Med. Chem.* **2008**, *16*, 6474.

93. Bate, A. B.; Kalin, J. H.; Fooksman, E. M.; Amorose, E. L.; Price, C. M.; Williams, H. M.; Rodig, M. J.; Cho, M. O.; Mitchell, S. H.; Wang, W.; Franzblau, S. G. *Bioorg. Med. Chem. Lett.* **2007**, *17*, 1346.

94. Darvesh, S.; Darvesh, K. V.; McDonald, R. S.; Mataija, D.; Walsh, R.; Mothana, S.; Lockridge, O.; Martin, E. *Eur. J. Med. Chem.* **2008**, *51*, 4200.

95. Kubota, K.; Kurebayashi, H.; Mayachi, H.; Tobe, M.; Onishi, M.; Isobe, Y. *Bioorg. Med. Chem. Lett.* **2009**, *19*, 2766.

96. Mitchell, S. C. *Curr. Drug Target* **2006**, *7*, 1181.

97. Okafor, C. O.; Okerelu, I. O.; Okeke, S. I. *Dyes Pigm.* **1986**, *8*, 11.

98. Pathak, V. N.; Yadav, S. S.; Srivastava, R. C. *Indian Sci. Abstr.* **1994**, *30*, 17.

99. Elliot, J. S.; Edwards; Brit. Pat. 860 (1961) 675, *Chem. Abstr.* **1961**, *55*, 15913.

100. Burford, B. L.; Karder, O. S. *Belz* 1964, *665*, 496; *Chem. Abstr.* **1966**, *64* 19902.

101. Mais, F. J.; Fiege, H. (Bayer A-G) Eur. Pat. Appl. EP505, 874(Cl CO7C17/12), 1992, DE Appl. 4, 110,057, 1991, 7pp, *Chem. Abstr.* **1993**, *118*, 22009.

102. De Bataafsche, N. V. *Brit. Pat.* **1958**, *789*, 947, *Chem. Abstr.* **1958**, *52*, 9585.

103. Gevaert Agfa, N. V. *Neth. Appl.* 1960, *6*, 605; *Chem. Abstr.* **1967**, *66*, 7121.

104. Shimizu, T.; Kaneko, I.; Shimakura, Y. *Eur. Pat.* **1983**, *126*, 39.

105. Mitchell, S. C. *Drug Metabol. Rev.* **1982**, *13*, 319.

106. Murphy, C. M.; Rarner, H.; Smith, N. L. *Ind. Eng. Chem.* **1950**, *42*, 2479.

107. Gerhard, M.; Siegfried, H.; Horst, W.; Toachim, L.; Manfred, N. *Ger. Pat.* **1979**, *139*, 269; *Chem. Abstr.* **1980**, *93*, 115943.

108. Smith, N. L. *U.S. Pat.* 2 **1952**, *287*, 661; *Chem. Abstr.* **1952**, *46*, 9182.

109. Foley, J. W.; Cincotta, L. U.S. 4, 962, 197, (Cl. 544-31; CO 7D 265/30), 1990; Appl. 157. 214, 10PP; *Chem. Abstr.* **1991**, *114*, 185524.

110. Motohashi, N.; Sasaki, Y.; Wakabayashi, Y.; Sakagami, H.; Molnar, J.; Kurihara, T. *Anticancer Res.* **1992**, *12*, 1423; *Chem. Abstr.* **1993**, *118*, 93805.

111. Cincotta, L.; Foley, J. W.; Eachern, T. M.; Lampros, E.; Cincotta, A. H. *Cancer Res.* **1994**, *54*, 1249.

112. Shukla, S.; Prakash, L. *Phosphorus Sulfur Silicon* **1995**, *102*, 39.

113. Aaron, J. J.; Maafi, M.; Kersebet, C.; Parkanyi, C.; Antonious, M. S.; Motohashi, N. *J. Photochem. Photobiol. Chem.* **1996**, *101*, 127.

114. Pusztai, R.; Motohashi, N.; Parkanyi, C.; Aaron, J. J.; Rao, B. K.; Molnar, J. *Anticancer Res.* **1996**, *16*, 2961.

115. Kurihara, T.; Watanabe, T.; Yoshikawa, K.; Motohashi, N. *Anti-cancer Res.* **1998**, *18*, 429.

116. Singh, G.; Swati; Mishra, A. K.; Prakash, L. *J. Fluorine Chem.* **1999**, *98*, 37.

117. Mucsi, I.; Molnar, J.; Motohashi, N. *Int. J. Antimicrob. Agents* **2001**, *18*, 67.

118. Ektova, L. V.; Bukhtoyarova, A. D.; Kolchina, E. F. In *3rd Euro Asian Heterocyclic Meeting "Heterocycles in Organic and Combinatorial Chemistry" (EAHM-2004)*.

119. Hilgeroth, A.; Molnár, A.; Molnar, J.; Voigt, B. *Eur. J. Med. Chem.* **2006**, *41*, 548.

120. Kumar, N.; Sharma, A. K.; Garg, R.; Yadav, A. K. *Indian J. Chem.* **2006**, *45B*, 747.
121. Kurihara, T.; Shinohara, K.; Inabe, M.; Wakabayashi, H.; Motohashi, N; Sakagami, H.; Molnar, J. *Bioactive Heterocycles VII Topics in Heterocyclic Chemistry*; 2008; Vol. 15, p 253.
122. Kang, W. B.; Nan'Ya, S.; Yamaguchi, Y.; Maekawa, E.; Ueno, Y. *J. Heterocycl. Chem.* **2009**, *24*, 91.
123. Aaron, J. J.; Gaye Seye, M. D.; Trajkovska, S.; Motohashi, N. *Heterocycl. Chem.* **2009**, *16*, 153.
124. Agarwal, N. L. Ph.D. Thesis, Rajasthan University: Jaipur, 1976.
125. Tandon, V. K.; Maurya, H. K.; Tripathi, A.; Keshava, G. B. S.; Shukla, P. K.; Srivastava, P.; Panda, D. *Eur. J. Med. Chem.* **2009**, *44*, 1086.
126. Onoabedje, E. A.; Ezema, B. E.; Ezema, C. G.; Ugwu, D. I. *Chem. Proc. Eng. Res.* **2013**, *8*, 6.
127. Ijeomah, A. O.; Okoro, U. C.; Godwin-Nwakwasi, E. U. *Int. J. Sci. Res.* **2014**, *12*, 449.
128. Ezema, B. E.; Ezema, C. G.; Ayogu, J. I.; Ugwu, D. I.; Onoabedje, E. A. *Orient. J. Chem.* **2015**, *31* (1), 379.
129. Ayuk, E. L.; Ilo, S. U.; Njokunwogbu, A. N.; Engwa1, G. A.; Oni, T. O.; Okoro, U. C. *Int. J. Mater. Chem.* **2015**, *5* (2), 44.
130. Fu, D. J.; Zhao, R. H.; Li, J. H. *Mol. Diversity* **2017**, *21*, 933.
131. Hu, W.; Zhang, S. *J. Org. Chem.* **2015**, *80* (12), 6128.
132. Burrell, A. K.; DelSesto, R. E.; Baker, S. N.; McCleskeya, S. N.; Bakerb, G. A. *Green Chem.* **2007**, *9*, 449.

CHAPTER 8

Investigations into the Interactions of Vinblastine with Proteins and Nucleic Acid

PRATEEK PANDYA[1] and NEELIMA GUPTA[2*]

[1]*Amity Institute of Forensic Sciences, Amity University, Noida, India*

[2]*Centre for Advanced Study, Department of Chemistry, University of Rajasthan, Jaipur, India*

Corresponding author. E-mail: guptaniilima@gmail.com; gupta_neelima@uniraj.ac.in

ABSTRACT

Vinblastine (VLB) is one of the alkaloids produced in substantial quantity by perennial shrub, *Vinca rosea*. Vinblastine is widely extracted from natural resources and used in chemotherapy of various types of hematological lymphoma and leukemia. Anticancer activity of VLB has been assigned mainly to its binding affinity to tubulin resulting in inhibition of tubulin oligomerization and cell apoptosis. Several methods have been used to study the binding interactions of VLB with tubulin and other biomolecules. UV spectroscopy, X-ray crystallography, and computational tools have been utilized to study the binding interactions. Herein, the methods available for the investigation of interaction of vinblastine with biomolecules such as proteins and nucleic acids and reasons for its anticancer activity have been discussed.

8.1 INTRODUCTION

Vinca alkaloids are obtained from the perennial shrub *Vinca rosea*, the latter also known as *sadabahar* in certain parts of India. *V. rosea* produces over 100 alkaloids; few of them are well known for their anticancer activity. Most commonly known bioactive vinca alkaloids include vinblastine (VLB), vincristine, vinorelbine, and vincamine. VLB is known to be mainly useful in the treatment of Hodgkin's disease, lymphocytic lymphoma, histiocytic lymphoma, advance testicular cancer, advanced breast cancer, Kaposi's sarcoma, etc. It is known to interfere with glutamic acid metabolism. The structural features of vinca alkaloids are studied using variety of instrumentation techniques, namely, IR spectroscopy, UV spectroscopy, mass spectrometry, NMR spectroscopy, circular dichroism studies, and X-ray crystallography.

8.2   CHARACTERIZATION OF VINBLASTINE

VLB is a bis-indole alkaloid, and its molecule has two distinct halves, namely, catharanthine and vindoline (Fig. 8.1). Structural features of VLB have been well characterized by comparing the IR spectra with catharanthine and vindoline for the presence of an indole moiety and a dihydroindole moiety (Gorman et al., 1959). VLB is commercially available as its sulfate salt.

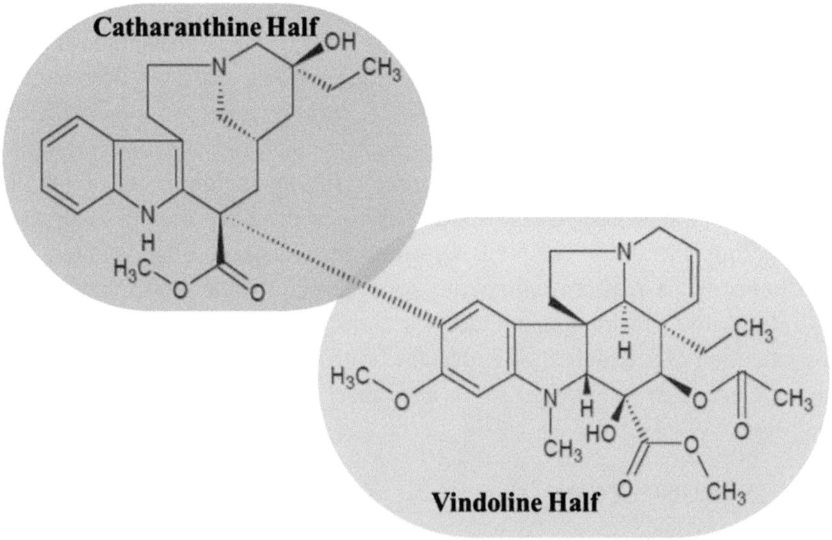

FIGURE 8.1   Catharanthine and vindoline halves of vinblastine (VLB) make it B/W.

VLB sulfate could be crystallized from Et_2O, and its reported (Neuss et al., 1959) melting point and specific rotation are [VLB (I) sulfate—mp, 284°–285°, $[\alpha]^{26}D$ –28° (MeOH)]. High-resolution mass spectrometry was used to identify the structure (Bommer et al., 1964) of VLB (VLB is $C_{46}H_{58}N_4O_9$, exact mass 810.4219) (Table 8.1), information regarding the attachment of the two parts was given by m/e = 509; the fragment of mass 154 ($C_9H_{16}NO$) was most abundant in vindoline half. Scanning thin layer chromatography (TLC) and TLC-UV techniques (Xianru et al., 1995) were employed in analyzing VLB using $CHCl_3$–MeOH–petroleum Ether (9:1:5) as mobile phase; VLB was determined by dual wavelength TLC scanner at 280 and 250 nm (Table 8.1).

TABLE 8.1 General Details of Vinblastine Alkaloid.

Common name	Vinblastine Vinblastine sulfate (salt)
IUPAC name	(2α,3α,4α,5β,19β)-Vincaleukoblastine
Molecular formula	$C_{46}H_{58}N_4O_9$
Molecular weight	810.989 g/mol
Canonical smiles	CCC1(CC2CC(C3=C(CCN(C2)C1) C4=CC=CC=C4N3)(C5=C(C=C6C(=C5) C78CCN9C7C(C=CC9)(C(C(C8N6C)(C(=O) OC) O)OC(=O)C)CC)OC)C(=O)OC)O
CAS number	865-21-4
No. of H-bond donor atoms	3
No. of H-bond acceptor atoms	12
Melting point	267°C
Dissociated constants	pKa1 = 5.4; pKa2 = 7.4
PubChem CID	13342

The UV spectrum (Kutney et al., 1975) of cleavamine ($C_{19}H_{24}N_2$) showed two characteristic bands at 280 and 225 nm. The vindoline, on the other hand, showed UV spectrum of characteristic benzenoid compounds (Kutney et al., 1975). This study indicated that the C18¢ center regulates the relative geometry of indole and indoline moieties. The electronic transitions of these two chromophores could be correlated to the stereochemistry at this center. Until the year 1975, the stereochemistry of C18¢ chiral center that links catharanthine (the indole chromophore cleavamine) and vindoline (indoline chromophore) units was not very well understood. To determine the stereochemistry at this location, CD (circular dichroism) and

X-ray studies were conducted. The CD studies of six naturally occurring and four synthetic dimers were carried out, which revealed that the evaluation of chirality of compounds can be studied by CD Technique (Kutney et al., 1968). There is sufficient theoretical basis available for correlation (Dong et al., 1995; Kutney et al., 1975) with experimental work. Absolute configuration of VLB in organic solvent was identified with NMR spectroscopy using ^{1}H, ^{13}C, COSY, DQF-COSY, and DQ Phase sensitive COSY experiments (Gaggelli et al., 1992). In addition, the conformation of VLB in aqueous solution was also determined by 2D ^{1}H–^{13}C NMR spectroscopy (Wenkert et al., 1973, 1975). Structure of VLB consists of well-defined planar and highly puckered regions (Fig. 8.2). Indole moiety in the catharanthine half is planar while rest of the catharanthine half is in chair form. Similarly, the indoline moiety in vindoline half is partially planar while the rest of the half is puckered.

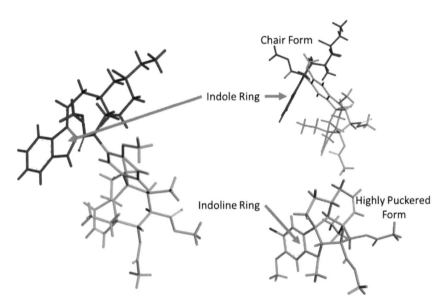

FIGURE 8.2   Different regions of flexibility in vinblastine molecule.

8.3   BINDING OF VINBLASTINE WITH BIOMOLECULES

Activity of VLB is due to the presence of several structural features making it capable to bind with biomolecules such as microtubules, HSA (human serum albumin), and DNA. In this chapter, the interaction of vinblastine with various biomolecules and methods used for the study of these interactions have been reviewed.

8.3.1 BINDING OF VINBLASTINE WITH TUBULIN PROTEIN

VLB is known to disrupt the self-assembly of microtubules by binding with beta subunit of tubulin protein, thereby hindering tubulin polymerization. VLB thus causes cell cycle arrest at metaphase/anaphase transition stage by disrupting spindle assembly (Kruczynski et al., 2002; Ngan et al., 2000, 2001). X-ray crystal studies have shown that the binding site on tubulin protein consists of 14 amino acid residues (Table 8.2) (Gigant et al., 2005).

TABLE 8.2 Binding Site Information of VLB–Tubulin Complex Crystal (Gigant et al., 2005).

S. No.	Residue	Name	Secondary structure
1.	PRO175	Proline	Coil
2.	LYS176	Lysine	Coil
3.	VAL177	Valine	Coil
4.	ASP179	Aspartic acid	Coil
5.	TYR210	Tyrosine	Helix
6.	PHE214	Phenylalanine	Helix
7.	THR220	Threonine	Coil
8.	THR221	Threonine	Coil
9.	PRO222	Proline	Coil
10.	THR223	Threonine	Coil
11.	TYR224	Tyrosine	Coil
12.	LEU227	Leucine	Helix
13.	PRO325	Proline	Helix
14.	VAL328	Valine	Helix
15.	ASN329	Asparagine	Helix
16.	PHE351	Phenylalanine	Coil
17.	VAL353	Valine	Sheet
18.	ILE355	Isoleucine	Sheet

A detailed analysis of binding interactions in VLB–tubulin crystal structure reveals that VLB, being larger in size than other conventional drugs, occupies an area that is significantly exposed to solvent environment. It is interesting to note that most of the residues in the binding site of VLB are less hydrophobic in nature. As shown in Table 8.3, variety of interactions, such as hydrogen bonding, hydrophobic contacts, and sigma-pi interactions are possible between VLB and tubulin at multiple locations.

TABLE 8.3   Intermolecular Noncovalent Bonding Pattern in VLB–Tubulin Complex.

S. No.	Name	Distance (Å)	Category	Type
1.	VLB:N9—B:ASP179:OD1	4.31682	Electrostatic	Attractive charge
2.	C:ASN329:ND2—VLB:N66	3.03248	Hydrogen bond	Conventional H-bond
3.	C:ASN329:ND2—VLB:O74	2.78904	Hydrogen bond	Conventional H-bond
4.	VLB:O72—B:PRO222:O	2.73516	Hydrogen bond	Conventional H-bond
5.	B:PRO222:CD—VLB:O75	3.42738	Hydrogen bond	Carbon hydrogen bond
6.	VLB:C55—B:VAL177:O	3.64167	Hydrogen bond	Carbon hydrogen bond
7.	VLB:C10—B:ASP179:OD1	3.09124	Hydrogen bond	Carbon hydrogen bond
8.	VLB:C8—C:PHE351:O	3.40709	Hydrogen bond	Carbon hydrogen bond
9.	VLB:C33—B:TYR210:OH	3.07652	Hydrogen bond	Carbon hydrogen bond
10.	C:PRO325:CH—VLB pi-orbital	3.35428	Hydrophobic	Pi-sigma
11.	C:PRO325 alkyl—VLB alkyl	4.9223	Hydrophobic	Alkyl
12.	VLB pi-orbital—C:PRO325 alkyl	4.2468	Hydrophobic	Pi-alkyl
13.	VLB pi-orbital—C:VAL328 alkyl	4.72703	Hydrophobic	Pi-alkyl
14.	VLB pi-orbital—C:VAL353 alkyl	4.41088	Hydrophobic	Pi-alkyl
15.	VLB pi-orbital—C:ILE355 alkyl	5.46985	Hydrophobic	Pi-alkyl

Note: Protein residues belonging to different chains are denoted as B or C.

8.3.2   BINDING WITH DNA

DNA is one of the most important targets in anticancer and antiviral therapy. Investigation of its interaction with various drug molecules has been reported in literature. The rationale that VLB could bind with DNA rests on the fact that it crosses the nuclear membrane and interacts with tubulin protein. Being so close to the nuclear DNA, the chances of its interaction with the DNA is very high. Earliest attempts toward identifying DNA binding were carried out by utilizing HPLC, UV spectroscopy, and DNA-cellulose column as a screening method for the assessment of DNA binding ability of VLB (Pezzuto et al., 1991). Experimental techniques such as UV–vis spectroscopy and IR-spectroscopy and computational methods such as molecular docking and QM–MM (quantum mechanics/ molecular mechanics) methods have established that VLB binds with duplex DNA, albeit with lesser affinity than tubulin protein.

There are several structural features present in VLB molecule that are thought to facilitate its binding with a variety of biological receptors. Features such as different regions of conformational flexibility, presence of H-bond donor, and acceptor groups are responsible for its binding activity. The apparent planarity of indole and indoline rings in VLB was initially perceived to help VLB molecule bind with DNA bases via minor groove site (Gupta et al., 2011; Pandya et al., 2010, 2012). However, FT-IR and UV–vis spectroscopy studies suggested that the binding of VLB to DNA is through A-T, G-C base pairs, and its phosphate backbone (Tyagi et al., 2012). It was also suggested that there could be a possibility of partial intercalation of VLB without disturbing the B-conformation of DNA. In the absence of a suitable model of VLB–DNA interaction, molecular modeling calculations were attempted to find out the structural basis of such interactions and provide a suitable model for VLB–DNA complex. Molecular docking and hybrid QM–MM calculations have suggested that major groove of DNA could be the best site for VLB binding (Fig. 8.3) as indicated by better binding affinity in major groove binding mode as compared to minor groove binding mode (Pandya et al., 2014). In addition, circular dichroism spectroscopy has revealed no conformational change in the DNA structure upon binding with VLB confirming results of FT-IR, major groove docking, and QM–MM calculations.

8.3.3 BINDING WITH HSA

HSA is the most abundant protein present in human blood. Its ability to bind with multiple ligands has made it one of the most studied carrier molecules. Drugs and other substances often travel to their destination by getting attached with HSA protein. This promiscuity of HSA is due to the presence of its well-defined regions (Domain I, II, and III) that are available for binding with multiple ligand molecules such as drugs, hormones, fatty acids, proteins (Ahmed-Ouameur et al., 2006; Ascenzi and Fasano, 2010; Ascenzi et al., 2006; Fasano et al., 2005; Gundry et al., 2007; Varshney et al., 2010). Computational investigation has indicated that VLB interacts substantially with HSA. Initial results from docking investigations (Fig. 8.4) estimated the binding possibility of VLB into the central part of HSA in the Tr-5 region (thyroxine binding site) that is skirted by domains I, II, and IIIA (Pandya et al., 2014).

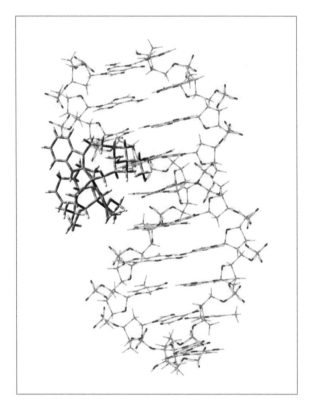

FIGURE 8.3   Binding of VLB in major groove of DNA.

8.3.4   *INTERACTIONS WITH OTHER BIOMOLECULES*

As the VLB is capable to cross the membrane barriers, it is highly likely that it may interact with other biomolecules as well. For instance, the investigations have revealed that it is likely to bind with nuclear receptors, such as NR1I2 (Smith et al., 2010), and multidrug efflux transporters, such as Pgp (Sharom et al., 1999). VLB has been found to bind with Pgp in two different sites with variable affinity (Mittra et al., 2017; Shapiro and Ling, 1997). Despite several such studies, a detailed structural characterization of binding interactions of VLB with other potential receptors is still not completely known and therefore is an area for further studies.

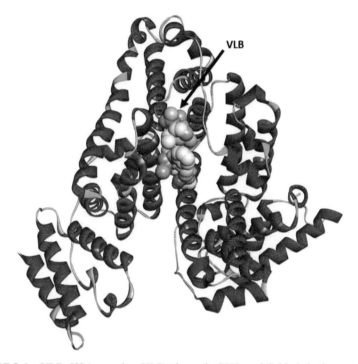

FIGURE 8.4 VLB–HSA complex. VLB (shown in CPK model) binds in the central Tr-5 region of HSA. Make it B/W.

KEYWORDS

- **vinblastine**
- ***Vinca rosea***
- **tubulin**
- **DNA**
- **drug–receptor interactions**
- **anticancer drug**

REFERENCES

1. Ahmed-Ouameur, A.; Diamantoglou, S.; Sedaghat-Herati, M. R.; Nafisi, S.; Carpentier, R.; Tajmir-Riahi, H. A. *Cell Biochem. Biophys.* **2006**, *45*, 203.
2. Ascenzi, P.; Bocedi, A.; Notari, S.; Fanali, G.; Fesce, R.; Fasano, M. *Mini Rev. Med. Chem.* **2006**, *6*, 483.
3. Ascenzi, P.; Fasano, M. *Biophys. Chem.* **2010**, *148*, 16.
4. Bommer, P.; McMurray, W.; Biemann, K. *J. Am. Chem. Soc.* **1964**, *86*, 1439.
5. Dong, J.-G.; Bornmann, W.; Nakanishi, K.; Berova, N. *Phytochemistry* **1995**, *40*, 1821.
6. Fasano, M.; Curry, S.; Terreno, E.; Galliano, M.; Fanali, G.; Narciso, P.; Notari, S.; Ascenzi, P. *IUBMB Life* **2005**, *57*, 787.
7. Gaggelli, E.; Valensin, G.; Stolowich, N. J.; Williams, H. J.; Scott, A. I. *J. Nat. Prod.* **1992**, *55*, 285.
8. Gigant, B.; Wang, C.; Ravelli, R. B.; Roussi, F.; Steinmetz, M. O.; Curmi, P. A.; Sobel, A.; Knossow, M. *Nature* **2005**, *435*, 519.
9. Gorman, M.; Neuss, N.; Svoboda, G. H. *J. Am. Chem. Soc.* **1959**, *81*, 4745.
10. Gundry, R. L.; Fu, Q.; Jelinek, C. A.; Van Eyk, J. E.; Cotter, R. J. *Proteomics: Clin. Appl.* **2007**, *1*, 73.
11. Gupta, S. P.; Pandav, K.; Pandya, P.; Kumar, G. S.; Barthwal, R.; Kumar, S. *Chem. Biol. Interact.* **2011**, *1*, 297.
12. Kruczynski, A.; Etievant, C.; Perrin, D.; Chansard, N.; Duflos, A.; Hill, B. *Br. J. Cancer* **2002**, *86*, 143.
13. Kutney, J. P.; Beck, J.; Bylsma, F.; Cretney, W. J. *J. Am. Chem. Soc.* **1968**, *90*, 4504.
14. Kutney, J. P.; Gregonis, D. E.; Imhof, R.; Itoh, I.; Jahngen, E.; Scott, A. I.; Chan, W. *J. Am. Chem. Soc.* **1975**, *97*, 5013.
15. Mittra, R.; Pavy, M.; Subramanian, N.; George, A. M.; O'mara, M. L.; Kerr, I. D.; Callaghan, R. *Biochem. Pharmacol.* **2017**, *123*, 19.
16. Neuss, N.; Gorman, M.; Svoboda, G.; Maciak, G.; Beer, C. *J. Am. Chem. Soc.* **1959**, *81*, 4754.
17. Ngan, V. K.; Bellman, K.; Hill, B. T.; Wilson, L.; Jordan, M. A. *Mol. Pharmacol.* **2001**, *60*, 225.
18. Ngan, V. K.; Bellman, K.; Panda, D.; Hill, B. T.; Jordan, M. A.; Wilson, L. *Cancer Res.* **2000**, *60*, 5045.
19. Pandya, P.; Agarwal, L. K.; Gupta, N.; Pal, S. *J. Mol. Graph. Model.* **2014**, *54*, 1.
20. Pandya, P.; Gupta, S. P.; Pandav, K.; Barthwal, R.; Jayaram, B.; Kumar, S. *Nat. Prod. Commun.* **2012**, *7*, 305.
21. Pandya, P.; Islam, M. M.; Kumar, G. S.; Jayaram, B.; Kumar, S. *J. Chem. Sci.* **2010**, *122*, 247.
22. Pezzuto, J. M.; Che, C.-T.; McPherson, D. D.; Zhu, J.-P.; Topcu, G.; Erdelmeier, C. A.; Cordell, G. A. *J. Nat. Prod.* **1991**, *54*, 1522.
23. Shapiro, A. B.; Ling, V. *Eur. J. Biochem.* **1997**, *250*, 130.
24. Sharom, F. J.; Liu, R.; Romsicki, Y.; Lu, P. *Biochim. Biophys. Acta—Biomembr.* **1999**, *1461*, 327.
25. Smith, N. F.; Mani, S.; Schuetz, E. G.; Yasuda, K.; Sissung, T. M.; Bates, S. E.; Figg, W. D.; Sparreboom, A. *Ann. Pharmacother.* **2010**, *44*, 1709.

26. Tyagi, G.; Charak, S.; Mehrotra, R. *J. Photochem. Photobiol., B.* **2012,** *108,* 48.
27. Varshney, A.; Sen, P.; Ahmad, E.; Rehan, M.; Subbarao, N.; Khan, R. H. *Chirality* **2010,** *22,* 77.
28. Wenkert, E.; Cochran, D. W.; Hagaman, E. W.; Schell, F.; Neuss, N.; Katner, A.; Potier, P.; Kan, C.; Plat, M. *J. Am. Chem. Soc.* **1973,** *95,* 4990.
29. Wenkert, E.; Hagaman, E. W.; Lal, B.; Gutowski, G. E.; Katner, A. S.; Miller, J. C.; Neuss, N. *Helv. Chim. Acta* **1975,** *58,* 1560.
30. Xianru, D.; Kunyi, N.; Hai, C.; Wenjia, C. *J. China Pharm. Univ.* **1995,** *3,* 157–159.

CHAPTER 9

Synthesis, Characterization, and Biological Evaluation of 12-Membered-N4-Macrocyclic Transition Metal Complexes

MONIKA UPADHYAY and NIGHAT FAHMI*

Department of Chemistry, University of Rajasthan, Jaipur 302004, India

Corresponding author. E-mail: nighat.fahmi@gmail.com; angel.molly123@gmail.com

ABSTRACT

The template condensation reaction between bis(diacetyl)-4-methyl-*o*-phenylenediamine with various diamines and metal salts in 1:1:1 resulted in mononuclear 12-membered tetraaza macrocyclic complexes [M = Co(II), Ni(II), Zn(II), and Cd(II)]. All complexes have been characterized by elemental analysis and various spectral techniques, that is, FT-IR, ^{1}H NMR, mass and electronic studies. An octahedral geometry has been proposed for all these complexes. On the basis of the IR and NMR spectral data, the involvement of azomethine nitrogen in coordination with the central metal ion has been confirmed. The in vitro antibacterial studies of the complexes have been recorded against some pathogenic bacteria and the results are quite promising.

9.1 INTRODUCTION

The chemistry of Schiff base is an important field in coordination chemistry.[1] Schiff bases are potential anticancer drugs and, when administered as their metal complexes, the anticancer activity of these complexes is enhanced in comparison to the free ligand.[2] This is due to their ability to react with a range of metal ions forming stable complexes that have applications in different fields.[3,4] A number of Schiff base complexes have been used as oxygen carriers to mimic complicated biological systems. The interest in the study of macrocyclic species based on transition metal compounds and multidentate ligands is an interesting field in chemistry and has been the subject of extensive research due to their physical properties, reactivity patterns, and potential applications in many important areas.[5] These multiple roles played by the naturally occurring macrocycle in biological systems is well known. The possibility of obtaining coordination compounds of unusual structure and stability is the reason for interest in the study of macrocyclic metal complexes. Macrocyclic metal complexes show high stability due to the number and type of donor atoms and their relative positions within the macrocyclic skeleton.[6] The selectivity for the extraction of metal ions by macrocyclic ligands that contain the combination of Oxo (O), Aza (N), Phospha (P), and Sulfa (S) depends on the flexibility and nature of the ligand backbone, nature, and number of donor atoms in addition to their relative position in the macrocyclic ligands and the cavity size of the macrocyclic core.[7,8] The template condensation methods play a significant role in macrocyclic chemistry[9] and the transition metal ions are used as the templating agent.[10] The chemistry of synthetic macrocycle compounds is also of great attention due to their use as pigments and dyes, magnetic resonance imaging (MRI) contrast agents,[11] and their biological activities, including antiviral, anticancer,[12–16] antifertile,[17] antimicrobial,[18–21] antitumor,[22] and antidibetic.[23] Macrocyclic complexes also play important roles because of their applications in catalytic systems.[24] Moreover, macrocyclic amides have potential applications in electrophosphorescence devices (EL) and homogenous catalysis.[25]

9.2 EXPERIMENTAL

All the chemicals used for the preparation of the ligands were of AR (Analytical Reagent) grade. All the diamines were purchased from Alfa Aesar. Solvents were distilled by appropriate drying agents and methods. All reactions were carried out under ambient conditions.

9.2.1 ANALYTICAL AND PHYSICAL MEASUREMENTS

The nitrogen and chlorine contents of the complexes were estimated by Kjeldahl's and Volhard's method, respectively.[26] Metals were estimated gravimetrically.[27] The Rast camphor method was used to determine the molecular weight of macrocyclic metal complexes. Melting point was determined by using capillaries in electrical melting-point apparatus. The electronic spectra of Co(II) and Ni(II) complexes were recorded on an ultraviolet–visible spectrophotometer 752/752N, infrared spectra of the ligands, and their complexes were recorded with the help of Shimadzu FTIR-550 spectrophotometer on KBr pellets. ^{1}H NMR spectra were recorded on a JEOL-DELTA2-NMR 400 MHz spectrometer in CDCl$_3$ by using TMS (tetramethylsilane) as the internal standard. The thin-layer chromatography (TLC) was used to check the purity of complexes.

9.2.2 PREPARATION OF LIGAND (ML)

The ligand (ML) was prepared by dissolving diacetyl (20 mmol, 1.72 g) in 40 mL of ethanol and then calculated amount of 3,4-diaminotoluene (10 mmol, 1.22 g) was added in 2:1 molar ratio. The reaction mixture was heated under reflux for 4–6 h on a ratio head. It was then concentrated to half of the volume. The solution was cooled and the excess solvent was removed by slow evaporation by keeping it in a desiccator overnight. The colored crystalline products so obtained were purified by recrystallization in the same solvent and dried in vacuo. The synthetic route for the ligand is shown in Scheme 9.1.

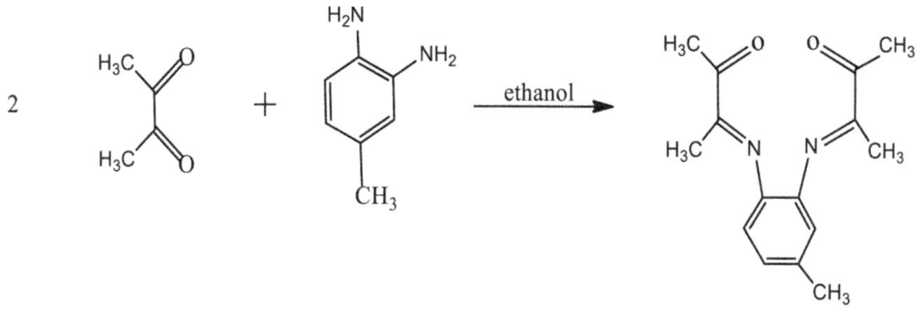

SCHEME 9.1 Synthetic route for the preparation of ligand.

9.2.3   PREPARATION OF METAL COMPLEXES

The ligand ML (10 mmol, 2.58 g) was dissolved in methanol and taken into a 100-mL round-bottom flask. Then the methanolic solution of diamines (4-fluoro-1,2-phenylenediamine [10 mmol, 1.26 g] and 4-chloro-1,2-phenylenediamine [10 mmol, 1.42 g]) and metal chlorides (10 mmol) was added to the solution of ligand in 1:1:1 molar ratio. After the addition of all the reagents, the mixture was boiled under reflux for 7–9 h on a ratio head. The volume of mixture was concentrated to its half and kept in a desiccator for 24 h at room temperature. The complexes obtained as solids were washed with methanol and dried under vacuo. The synthetic route for the preparation of metal complexes was showed in Scheme 9.2.

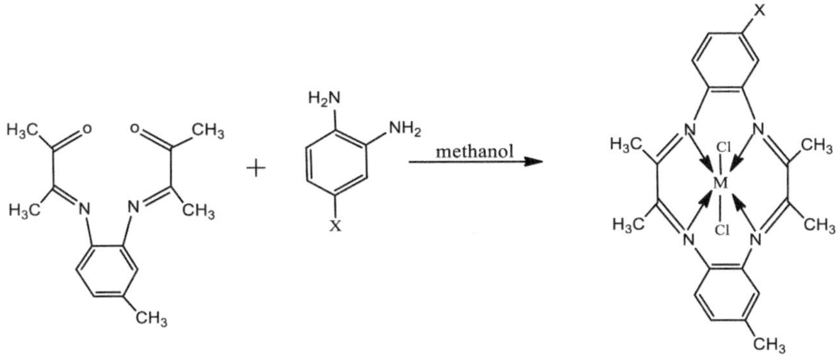

Where X= Cl and F
M= Co, Ni, Zn and Cd

SCHEME 9.2   The synthetic route for the preparation of metal complexes.

9.3   BIOLOGICAL ASSAY

9.3.1   TEST MICROORGANISM

Antibacterial properties of all complexes were evaluated and compared with the standard drug streptomycin. The microorganisms used were *Escherichia coli, Staphylococcus aureus, Pseudomonas aeruginosa,* and *Bacillus subtilis.*

9.3.2 IN VITRO ANTIBACTERIAL ACTIVITY

All the test compounds were dissolved in DMSO (dimethyl sulfoxide) and the concentration of stock solution was made 5 mg/mL. Each well was filled with 50 μL of test compound. Streptomycin was used as positive control (5 mg/mL concentration). Agar plates were prepared for the antibacterial activity. Sabouraud dextrose agar medium and Mueller–Hinton agar medium are susceptibility test media that have been validated by CLSI (Clinical and Laboratory Standards Institute) for screening the antibacterial activity by disk/well diffusion susceptibility testing. Fresh cultures of test isolates of *E. coli, S. aureus, P. aeruginosa,* and *B. subtilis* were inoculated in peptone water and kept for incubation for 30 min at 37°C. The bacterial suspensions were compared to 0.5 McFarland turbidity standard. Bacterial cultures were swabbed onto the Mueller–Hinton agar surface. Wells were then filled with 50 μL of different dilutions prepared from stock. The *E. coli, S. aureus, P. aeruginosa,* and *B. subtilis* plates were kept for incubation at 37°C for 24–48 h and results were observed.

The same procedure was adopted for screening the antibacterial activity of standard antibiotic streptomycin. The medium with DMSO as solvent was used as a negative control, whereas medium with streptomycin was used as a positive control. The experiments were repeated for three times.

9.4 RESULTS AND DISCUSSION

The tetraaza Schiff base macrocyclic complexes of Co(II), Ni(II), Zn(II), and Cd(II) have been synthesized by [2+2] condensation reaction. The resulting macrocyclic complexes are colored, solids, stable at room temperature. They are soluble in DMF (dimethylformamide), acetonitrile, $CDCl_3$, and DMSO. Molecular-weight determination showed monomeric nature of the complexes. The purity of the complexes was checked by TLC run in 1:1 benzene–methanol. The elemental analyses for all the complexes are shown in Table 9.1.

9.4.1 IR SPECTRA

In IR spectra of 3,4-diaminotoluene, two bands are observed at 3350 and 3390 cm^{-1}, corresponding to the $v(NH_2)$ group, and were absent in

TABLE 9.1 The Physicochemical Properties and Analytical Data of Ligand and Its Complexes.

S. No	Compounds	Color	Yield (%)	Mol. weight (g/mol)	Elemental analysis found (calculated) (%)				
					C	H	N	Cl	M
1.	[C$_{15}$H$_{18}$N$_2$O$_2$]	Dark brown	82	258.22 (258.31)	69.64 (69.74)	6.87 (7.02)	10.67 (10.84)	–	–
2.	[C$_{21}$H$_{21}$N$_4$Cl$_3$Co]	Brown	62	494.65 (494.70)	50.71 (50.98)	4.15 (4.27)	11.15 (11.32)	21.34 (21.49)	11.79 (11.91)
3.	[C$_{21}$H$_{21}$N$_4$Cl$_2$FCo]	Black	70	478.19 (478.25)	52.57 (52.73)	4.29 (4.42)	11.58 (11.71)	14.77 (14.82)	12.19 (12.32)
4.	[C$_{21}$H$_{21}$N$_4$Cl$_3$Ni]	Brown	63	494.30 (494.46)	50.86 (51.00)	4.14 (4.28)	11.17 (11.33)	21.32 (21.50)	11.66 (11.86)
5.	[C$_{21}$H$_{21}$N$_4$Cl$_2$FNi]	Dark brown	71	477.98 (478.01)	52.60 (52.76)	4.36 (4.42)	11.60 (11.72)	14.71 (14.83)	12.09 (12.27)
6.	[C$_{21}$H$_{21}$N$_4$Cl$_3$Zn]	Dark brown	68	501.07 (501.16)	50.20 (50.32)	4.07 (4.22)	11.01 (11.17)	21.15 (21.22)	12.91 (13.04)
7.	[C$_{21}$H$_{21}$N$_4$Cl$_2$FZn]	Black	72	484.65 (484.71)	48.40 (48.48)	3.93 (4.06)	10.59 (10.77)	20.30 (20.44)	12.42 (12.57)
8.	[C$_{21}$H$_{21}$N$_4$Cl$_3$Cd]	Dark brown	65	548.08 (548.18)	45.93 (46.01)	3.69 (3.86)	10.06 (10.22)	19.23 (19.40)	20.37 (20.50)
9.	[C$_{21}$H$_{21}$N$_4$Cl$_2$FCd]	Black	69	531.62 (531.73)	44.34 (44.46)	3.65 (3.73)	9.74 (9.87)	17.60 (18.75)	19.68 (19.81)

the infrared spectra of all the macrocyclic complexes. Moreover, no strong absorption band was observed near 1716 cm⁻¹ and this indicates the absence of the C=O group of the diacetyl moiety. The disappearance of these bands and the appearance of a new strong absorption band in the range 1610–1630 cm⁻¹ confirm the condensation of the carbonyl group of diacetyl and the amino group of diamines, and these bands may be assigned to v(C=N) stretching vibrations. The lower value of the v(C=N) vibrational bands can be explained by the shifting of electron density of the azomethine nitrogen toward the metal atom, and this indicates that the coordination occurs through the nitrogen of the C=N groups. In the region 2850–2950 cm⁻¹, the medium intensity bands were present, which may be assigned to the v(C–H) stretching vibrations of the methyl group of the diacetyl moiety. The v(C=C) aromatic stretching vibrations of the benzene ring show various absorption bands in the region 1400–1575 cm⁻¹. The band in the region 720–780 cm⁻¹ corresponds to the v(C–H) out of plane bending of the aromatic ring. The far infrared spectra show bands in the region 430–460 cm⁻¹, which corresponds to the v(M–N) vibrations and identify coordination of the azomethine nitrogen.[28]

9.4.2 ¹H NMR

¹H NMR spectra of the synthesized compounds were recorded in DMSO solution, using TMS as an internal standard.

In the ¹HNMR spectra, aromatic protons of ligand showed a multiplet at 7.19–7.25 ppm and methyl groups attached to C=N and C=O bonds showed two singlets that appeared at 2.07 and 2.30 ppm, respectively. The methyl protons of 3,4-diaminotoluene also show a singlet at 2.34 ppm.

The NMR spectra of the complexes confirm the formation of fully condensed macrocyclic Schiff base complexes. The ¹H NMR spectrum of the zinc(II) and Cd(II) complexes shows multiplets in the region 6.95–7.85 ppm corresponding to aromatic ring protons. A sharp signal in the range of 2.34–2.48 ppm may be assigned to the methyl protons of 3,4-diaminotoluene and another singlet appeared in the range 2.07–2.35 ppm, correspond to methyl protons of diacetyl moiety (Table 9.2).

TABLE 9.2 ^{1}H NMR Spectral Assignments for Zn(II) and Cd(II).

S. No.	Compounds	^{1}H NMR in DMSO-d$_6$ (δ, ppm)			
		Methyl protons attached to C=N	Methyl protons attached to C=O	Methyl protons of 3,4-diaminotoluene moiety 3H (s)	Aromatic ring protons (m)
1.	$[C_{15}H_{18}N_2O_2]$	2.07	2.30	2.34	7.19–7.25
2.	$[C_{21}H_{21}N_4Cl_3Zn]$	2.34	–	2.48	7.49–7.81, 7.40–7.54
3.	$[C_{21}H_{21}N_4Cl_2FZn]$	2.31	–	2.46	6.95–7.58, 7.42–7.59
4.	$[C_{21}H_{21}N_4Cl_3Cd]$	2.30	–	2.45	7.52–7.85, 7.42–7.56
5.	$[C_{21}H_{21}N_4Cl_2FCd]$	2.28	–	2.42	6.97–7.60, 7.45–7.63

Note: Where s, singlet; m, multiplet.

9.4.3 MASS SPECTRA

The mass spectra of macrocyclic compounds showed molecular ion peaks (M^{+1}) at m/z 477.02 and 476.93 for [$C_{21}H_{21}N_4Cl_2FCo$] and [$C_{21}H_{21}N_4Cl_2FNi$], respectively, which are in good agreement with the respective molecular formula. The spectra of complexes also show the various peaks due to fragment ions. The spectrum of [$C_{21}H_{21}N_4Cl_2FCo$] complex shows peaks at 348.02, 317.09, 297.15, 281.18, and 147.10 for [$C_{21}H_{21}N_4F$]$^+$, [$C_{14}H_{15}N_4FCo$]$^+$, [$C_{14}H_{15}N_4Co$]$^+$, [$C_{13}H_{10}N_4Co$]$^+$, and [$C_7H_8N_4$]$^+$ ions, respectively. The spectrum of [$C_{21}H_{21}N_4Cl_2FNi$] complex shows peaks at 402.93, 398.93, 180.87, and 176.88 for [$C_{16}H_6N_4FCl_2Ni$]$^+$, [$C_{17}H_9N_4Cl_2Ni$]$^+$, [$C_{10}H_3N_4$]$^+$, and [$C_{10}H_9N_2F$]$^+$ ions, respectively. This confirms the formation of the macrocyclic frame and also these data confirm the mononuclear nature of the Zn(II) complexes. The mass spectra of the complexes are presented in Figures 9.1 and 9.2.

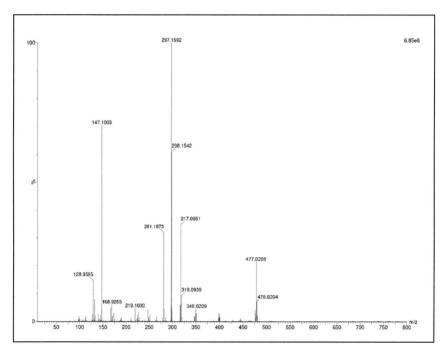

FIGURE 9.1 Mass spectra of [$C_{21}H_{21}N_4Cl_2FCo$] complex.

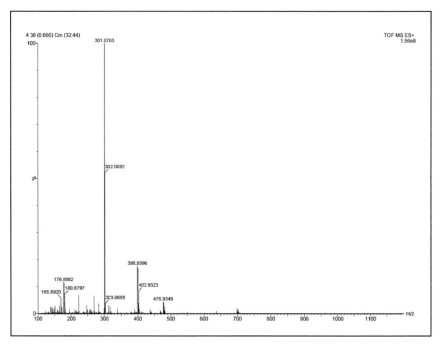

FIGURE 9.2   Mass spectra of $[C_{21}H_{21}N_4Cl_2FNi]$ complex.

9.4.4 ELECTRONIC SPECTRA

The electronic spectra of the complexes were recorded in DMSO in order to assign the geometries around the metal ions. The electronic spectra of the Co(II) complexes show three spins allowed bands at 8000, 14,200, and 21,000 cm^{-1} corresponding to $^4T1g(F) \rightarrow {}^4T1g(F)$, $^4T1g(F) \rightarrow {}^4A2g(F)$, and $^4T1g(F) \rightarrow {}^4T1g(P)$ transitions, and the electronic spectrum of Ni(II) complexes exhibits three bands at 8500, 15,000, and 22,200 cm^{-1} attributed to three spins allowed transitions viz. $^3A2g(F) \rightarrow {}^3T2g(F)$, $^3A2g(F) \rightarrow {}^3T1g(F)$, and $^3A2g(F) \rightarrow {}^3T1g(P)$. The v_2/v_1 was also calculated and is in close agreement with the octahedral geometry around the metal ions, which is further supported by magnetic moment value of 4.49 B.M. for Co(II) complexes and the observed magnetic moment value of 3.12 B.M. for Ni(II) complexes further complements the electronic spectral findings.

9.4.5 ANTIBACTERIAL ACTIVITY

The antibacterial activities of all the complexes were determined against four bacterial strains, that is, *E. coli*, *S. aureus*, *P. aeruginosa*, and *B. subtilis* then compared with the standard drug streptomycin.

All the complexes were found to be active on all types of bacterial strains. The antibacterial studies show an enhancement after the coordination of ligand with the metal ions. The polarity of the metal ion is reduced on complexation by the electron delocalization over the macrocyclic ring and the partial sharing of positive charge of metal ion to the donor groups present in the macrocyclic complex.[29] The increased activity after complexation may also because the chelation Effect. Chelation will augment the lipophilic character of the central metal atom which afterward favors its permeation through the lipid layers of the cell membrane and blocking the metal binding sites on enzymes of microorganisms.[30] The results of the antibacterial activity are summarized in Table 9.3 and the comparative results for antibacterial studies are shown in Figure 9.3.

TABLE 9.3 Inhibition Zones (mm) of Complexes against Bacterial Strains.

S. No.	Compounds	Diameter of inhibition zone (mm)			
		Bacillus subtilis	Staphylococcus aureus	Pseudomonas aeruginosa	Escherichia coli
1.	$[C_{15}H_{18}N_2O_2]$	8.2 ± 0.13	8.0 ± 0.18	8.7 ± 0.12	7.7 ± 0.10
2.	$[C_{21}H_{21}N_4Cl_3Co]$	13.0 ± 0.07	13.7 ± 0.15	16.1 ± 0.02	15.0 ± 0.18
3.	$[C_{21}H_{21}N_4Cl_2FCo]$	13.2 ± 0.11	14.1 ± 0.05	16.5 ± 0.02	15.6 ± 0.18
4.	$[C_{21}H_{21}N_4Cl_3Ni]$	11.0 ± 0.09	14.4 ± 0.07	15.3 ± 0.02	14.2 ± 0.18
5.	$[C_{21}H_{21}N_4Cl_2FNi]$	11.4 ± 0.03	14.8 ± 0.08	15.7 ± 0.02	14.8 ± 0.18
6.	$[C_{21}H_{21}N_4Cl_3Zn]$	14.5 ± 0.08	15.0 ± 0.13	13.1 ± 0.02	16.5 ± 0.18
7.	$[C_{21}H_{21}N_4Cl_2FZn]$	14.8 ± 0.06	15.2 ± 0.02	12.8 ± 0.02	16.8 ± 0.18
8.	$[C_{21}H_{21}N_4Cl_3Cd]$	7.8 ± 0.12	9.7 ± 0.09	8.5 ± 0.02	11.0 ± 0.18
9.	$[C_{21}H_{21}N_4Cl_2FCd]$	8.0 ± 0.04	10.4 ± 0.07	9.0 ± 0.02	11.3 ± 0.18
10.	Streptomycin	16.0 ± 0.01	18.0 ± 0.11	28.0 ± 0.02	25.0 ± 0.18

9.5 CONCLUSION

We report the successful synthesis of 12-membered macrocyclic complexes of Co(II), Ni(II), Zn(II), and Cd(II) metals by the template condensation.

The complexes have been characterized by a variety of physicochemical and spectroscopic methods, and on the basis of these studies, an octahedral geometry has been proposed for all the macrocyclic complexes. Also, antibacterial studies showed that all the prepared complexes have relatively antibacterial effects against the studied bacterial strains. The enhanced activity of the macrocyclic complexes than the parent ligand has been explained on the basis of chelation theory.

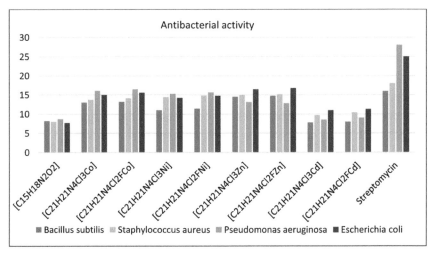

FIGURE 9.3 Antibacterial activities of ligands and their metal complexes with different bacteria.

KEYWORDS

- **tetraaza macrocycles**
- **Schiff base**
- **template condensation**
- **spectral studies**
- **antibacterial studies**

REFERENCES

1. Shiu, K. B.; Liu, S. A.; Lee, G. H. *Inorg. Chem.* **2010,** *49* (21), 9902.
2. Shahabadi, N.; Kashanian, S.; Darabi, F. *Eur. J. Med. Chem.* **2010,** *45*, 4239.
3. Yeşilel, O. Z.; Erer, H.; Kaştas, G.; Kani, I. B. *Polyhedron* **2010,** *29* (13), 2600.
4. Habib, H. B.; Gil-Hernández, B.; Abu-Shandi, K.; Sanchiz, J.; Janiak, C. *Polyhedron* **2010,** *29* (12), 2537.
5. Belghoul, B.; Weiterlich, I.; Maier, A.; Toutianoush, A.; Raman Rabindranath, A.; Tieke, B. *Langmuir* **2007,** *23* (9), 5062.
6. Ilhan, S.; Temel, H.; Yilmaz, I.; Sekerci, M. *J. Organomet. Chem.* **2007,** *692*, 3855.
7. Fernández-Fernández, M. C.; Bastida, R.; Macías, A.; Valencia, L.; Pérez-Lourido, P. *Polyhedron* **2006,** *25*, 783.
8. Temel, H.; Ilhan, S. *Spectrochim. Acta A: Mol. Biomol. Spectrosc.* **2008,** *69* (3), 896.
9. Niasari, M. S.; Daver, F. *Inorg. Chem. Commun.* **2006,** *9*, 175.
10. Prasad, R. N.; Gupta, S.; Jangir, S. *J. Indian Chem. Soc.* **2007,** *84*, 1191.
11. Chandra, S.; Pundir, M. *Spectrochim. Acta A* **2008,** *69*, 1.
12. Illan-Cabeza, N. A.; Hueso-Urena, F.; Moreno-Carretero, M. N.; Martinez-Martos, J. M.; Ramirez-Exposito, M. J. *J. Inorg. Biochem.* **2008,** *102*, 647.
13. Kumar, G.; Kumar, D.; Devi, S.; Johari, R.; Singh, C. P. *Eur. J. Med. Chem.* **2010,** *45*, 3056.
14. El-Boraey, H. A.; Serag El-Din, A. A. *Spectrochim. Acta A: Mol. Biomol. Spectrosc.* **2014,** *132*, 663.
15. EL-Boraey, H. A.; EL-Gammal, O. A. *Spectrochim. Acta A* **2015,** *138*, 533.
16. El-Boraey, H. A.; El-Salamony, M. A.; Hathout, A. A. *J. Incl. Phenom. Macrocycl. Chem.* **2016,** *86*, 153.
17. Costamagna, J.; Ferraudi, G.; Matsuhiro, B. *Coord. Chem. Rev.* **2000,** *196* (1), 125.
18. Singh, D.; Kumar, K. *J. Serb. Chem. Soc.* **2010,** *75* (4), 475.
19. Rajakumar, P.; Sekar, K.; Shanmugaiah, V.; Mathivanan, N. *Eur. J. Med. Chem.* **2009,** *44*, 3040.
20. Biswas, F. B.; Roy, T. G.; Rahman, Md. A.; Emran, T. B. *Asian Pac. J. Trop. Med.* **2014,** *7* (1), S534.
21. Stella Shalini, A. S.; Amaladasan, M. *Int. J. Innov. Res. Sci. Eng. Technol.* **2016,** *5* (2), 2053–2059.
22. El-Boraey1, H. A.; EL-Gammal, O. A. *Open Chem. J.* **2018,** *5*, 51.
23. Sakurai, H.; Kojima, Y.; Yoshikawa, Y.; Kawabe, K.; Yasui, H. *Coord. Chem. Rev.* **2002,** *226*, 187.
24. Singh, D. P.; Kumar, R.; Malik, V.; Tyagi, P. *Trans. Met. Chem.* **2007,** *32*, 1051.
25. Mohamed, A. A. A. *Coord. Chem. Rev.* **2010,** *254*, 1918.
26. Vogel, A. I. *A Text Book of Inorganic Analysis*; Longmans Green and Co: London, 1968.
27. Vogel, A. I. *A Text Book of Quantitative Inorganic Analysis*; Longmans Green, ELBS: London, 1962.
28. Gupta, L. K.; Chandra, S. *Spectrochim. Acta A* **2008,** *71*, 496.
29. Arslan, H.; Duran, N.; Borekci, G.; Ozer, C. K.; Akbay, C. *Molecules* **2009,** *14*, 519.
30. Chouhan, Z. H.; Hanif, M. *App. Organ. Chem.* **2011,** *25*, 753.

Index